The Sea Can Wash Away All Evils

Kimberley C. Patton

The Sea Can Wash Away All Evils

Modern Marine Pollution and the Ancient Cathartic Ocean

COLUMBIA UNIVERSITY PRESS NEW YORK

Columbia Unversity Press
Publishers Since 1893
New York, Chichester, West Sussex

Library of Congress Cataloging-in-Publication Data

Patton, Kimberley C. (Kimberley Christine)
   The sea can wash away all evils : modern marine pollution and the ancient
Cathartic Ocean / Kimberly C. Patton.
      p.   cm.
   Includes bibliographical references and index.
   ISBN 13: 978-0-231-13806-2 (cloth : alk. paper)
              978-0-231-51085-1 (e-book)
      1. Water—Religious aspects.   2. Ocean—Religious aspects.   3. Ocean.
4. Marine pollution.   5. Purity, Ritual.
      1. Title
BL450 .P38   2007
201/.77—DC22   2006023553

Columbia University Press books are printed on permanent and durable acid-free paper
Printed in the United States of America

c 10 9 8 7 6 5 4 3 2

References to Internet Web Sites (URLs) were accurate at the time of writing. Neither the author nor Columbia University Press is responsible for Web sites that may have expired or changed since the book was prepared

Permission of the publisher or current holder of copyright is gratefully acknowledged to reproduce the following:

Mary Oliver, "The Sea," from *American Primitive* (Little, Brown, 1983); "The Mother of the Sea Beasts," from *Traditional Stories of Eskimo and Indian People,* ed. Howard Norman (Pantheon Books, 1990), 212–214 (reprinted with permission of Melanie Jackson Agency, LLC); *The Śiva Purāṇa* 2.3.20, 1–23, trans. by a board of scholars (Delhi: Motilal Banarsidass, 1973–1974); *Subhāṣitaratnakoṣā* 1196, 1198, 1209, from *An Anthology of Sanskrit Court Poetry: Vidyākara's "Subhāṣitaratnakoṣā,"* trans. Daniel Ingalls, Harvard Oriental Series vol. 44 (Cambridge: Harvard University Press, 1965), pp. 338–339, 341 copyright© 1965 by the President and Fellows of Harvard College.

*For my mother, Christine, who taught me to love the sea,
and for the parents bereaved by the Indian Ocean
who could not save their children from it*

December 26, 2004

θάλασσα κλύζει πάντα τἀνθρώπων κακά
"The sea can wash away all evils of humankind"

—Euripides, Iphigenia in Tauris 1193

Contents

Preface

Years ago when I began work on this book, inspired by Euripides' tragedy *Iphigenia in Tauris,* by the wild Inuit sea-spirit Sedna, and by the disquieting image of the submarine mare-fire in the Hindu Purāṇas, I tried to understand why human religious thought and practice have so often turned to the sea as a way of staying "free." In these stories, the sea can wash away or absorb what otherwise would be unbearable if allowed to remain on land, the normal theater of ritual life. The sea's ancient role as the ultimate place of catharsis is prominent in many religious traditions. It makes terrestrial existence possible by keeping it "pure." Historically, many cultures have revered the sea, and at the same time they have made it to bear and to wash away whatever was construed as dangerous, dirty, or morally contaminating.

I have long wondered whether modern marine pollution, now threatening the sea in immeasurable ways, may stem from similar responses. I locate the genesis of the contemporary habits and attitudes of industrialized nations toward the World Ocean in the sea's own natural qualities: its vast size and depth; its chronic motion in currents, tides, and waves; its apparent biotic inexhaustibility; its mystery, antiquity, and chaos; and its composition of traditional purifiers, salt and water. Despite increasing evidence to the contrary, most human beings do not really grasp that anything we might do to the oceans might harm them. In this book, I argue that environmental science—and environmental advocacy—ignores the history of religion at a cost, since the collective impulses resulting in ecological damage are the effects of larger and older reflexes. In other words, even when highly secularized, cultural habits may be deconstructed in terms that I believe share some of the same assumptions as do religious patterns of thought. Both accord agency to the sea while at the same time treating it as a passive receptacle. The ocean has been seen as both lethal and life-giving to human beings, many of whom have lived in coastal societies. This book is about the history of our ambivalence toward the ocean, through the dichotomy of pollution and purification, and tracks the consequences of that history in both religious chronicles and the projections of marine ecology.

As I worked to complete this project, another powerful dimension of the ocean rose up, demanding its place in these pages, and reconfiguring the sea's own cultural archetypes. In the early morning of December 26, 2004, the sea brought disaster to South and Southeast Asia off the western coast of Indonesia when the Indian Plate ground itself beneath the edge of the Burmese Plate and snapped it, creating a monstrous wave. Traveling five hundred miles an hour along the floor of the Indian Ocean, a tsunami—as high as fifty feet near the epicenter of the undersea quake—surfaced along coastlines, in some places many hours later, and swamped island chains from Sumatra to Thailand to the Andaman and Nicobar Islands; it hit populations from Bangladesh to the eastern coast of India to Sri Lanka to the Maldives to Madagascar. The waves drowned at least 230,000 people, half of them children.

In India, the tsunami came on the auspicious day of the full-moon ceremony, when the tides are higher. At this time of the month, Hindus traditionally wade into the sea to pay homage to the celestial bodies, just as Brahmins purify themselves by walking into water every morning and reciting the *gāyatrī mantra* to the sun. But this time, instead of a purifying communion with the water, there came a merciless wall of death. For those along thousands of miles of coast-line, the immediate world was swept away in a matter of seconds. Houses, temples, and churches were flattened; trains thrown over; boats shattered; villages destroyed. Babies and children were torn from their parents' arms by the force of the waves, some of them heartlessly borne back again as tiny corpses, thrown up on the sand or left in trees. Many were lost forever in the boiling surf.

I mention the reconfiguration of the sea as a cultural entity in the wake of this event. It is also a theological reconfiguration, hideously wrought, the eventual shape of which cannot be known now. The reconfiguration began instantly and is both highly specific to the societies directly affected, and also international and cross-cultural in scope. Because the Indian Ocean has not suffered a tsunami in historical memory, the sea there, for the most part, had no mythical presence as a dealer of death[1] apart from medieval Hindu apocalyptic, which is discussed in chapter 6. On the contrary, the sea was generally not seen as a place of death to be avoided in the coastal cultures it devastated but as a soothing omnipresence, a source of beauty and livelihood through fishing and tourism. But now, in the words of Sri Lankan fisherman Subakean Albino, born in a house by the sea seventy years ago, "We see it as a mother. Our mother has punished us."[2] In the large and small fish markets of Indonesia and coastal Thailand, sales dropped dramatically, despite many who were hungry, for fear that freshly caught fish had fed on human remains.

On January 5, 2005, Amy Waldman and David Rohde published an article in the *New York Times* about the radical shift in popular attitudes toward the sea in the tsunami-hit areas of Sri Lanka. For example, R. G. Jayadsara of Hambantota said that when he looked out to sea, once a source of life, he now saw a cemetery. "He pointed to where the tsunami had rewritten the coastline, tracing curves in place of a straight edge. It seemed to reflect the way the murderous surge also rewrote, perhaps permanently, the covenant between the people of this island nation and the sea that surrounds it." In Matara, whose population "until December 26 liked nothing better than a sea bath on a Sunday," Dudley Silva, an irrigation engineer, observed, "Now people hate the sea—they hate it." He told of seeing a woman standing by the ocean a few days after the wave, waving her arms and cursing it.[3]

On the northern coast, in the village of Mullaittivu, Hindus worship the sea as a life-giving goddess, and Christians dab seawater on their foreheads and eyelids, invoking the Holy Mother and St. Anthony to protect themselves against shipwreck.[4] When Tamil rebels caused Selva Malar and her family to flee Mullaittivu to India for a period, she remembered, "We used to go to a church several miles away, near the shore, just to feel like it was home." The presence of the ocean was essential to a sense of orientation, a human sense with its often unacknowledged geophysical element, just as those raised in the mountains grow restless when compelled to live far from them. Indra Gamaga, a Buddhist whose temple provided sanctuary for hundreds, says that she will leave and never return to live near the sea. "The image of so many bodies had lodged in her mind, transposed with the image of the sea."[5]

On December 26, as Rev. Charles Hewawasam was giving Holy Communion at Our Lady of Matara Church on the southern Sri Lankan coast, his church was engulfed by the sea. Shouting to warn his congregants, he fled to the top of a new building next door, returning to try to save the eighteen-inch statue of Mary and the infant Jesus, already swept away before another wave slammed the church. The nun giving communion next to him died, along with eighteen congregants.

The statue, too, survived miraculously. History, or legend, has it that the statue first came from the sea, when it was found by fishermen 500 years ago. It was lost twice after that, once in a ship wreck en route to Europe to be painted and again on the way back, when it was misplaced. It was found both times, and again now, three days after the tsunami, in a garden. "She came from the sea," Mr. Hewawasam said. "She knows how to swim."[6]

Such stories of the miraculous, along with those that are tragic and seem to bespeak Heaven's cruelty, indifference, or lack of existence, will surely proliferate as the collective memory of this event is shared and religiously shaped by those who experienced it. What Holocaust scholar Nehemia Polen has called "the mysticism of catastrophe" is a generator of meaning to both those in the grip of the catastrophe and those who witness their suffering. With the photographs, films, and reports of the 2004 tsunami so rapidly disseminated around the world, these witnesses are in the billions. The world's populations were able—indeed, compelled—to engage with the sea's unexpected new persona in the Indian Ocean. The waves inundated some of the most beautiful and verdant beaches in the world, magnetic to tourists and a major source of local industry, as well as an indicator of the great economic disparities between local and foreign populations. The sea left the Edenic beaches as desolate as charnel houses, even as, from an ecological viewpoint, it restored Eden by wiping away human habitus. Perhaps the tsunami also also enacted, in waking life, a nightmare that has come to many at many times and places, one that recurred in the sleep of J. R. R. Tolkien throughout his life, which he gave to the character of Faramir of Gondor in *The Return of the King*:

A sound like a sigh went up from all the lands about them; and their hearts beat suddenly again.

"It reminds me of Númenor," said Faramir, and wondered to hear himself speak.

"Of Númenor?" said Eowyn.

"Yes," said Faramir, "of the land of Westernesse that foundered. And of the great dark wave climbing over the green lands and above the hills, and coming on, darkness inescapable. I often dream of it."[7]

We are grappling with the sea's conversion from paradise to hell. Already the sea is being globally re-imagined in the wake of the greatest tsunami ever to strike in recorded history. In the Americas, this shift was reiterated in the wake of repeated monster hurricanes with their attendant marine phenomena striking the coastal Gulf and Caribbean islands in the summer and fall of 2005. Less spasmodically but no less threateningly, accelerated global warming has caused the average sea level to rise 10–20 cm (4–8 in.) during the twentieth century as polar ice melts, "and it is expected to rise a further 25–100 cm (10–40 in.) in the first few years of the twenty-first century as melting continues, not least because seawater expands slightly as it warms."[8] The ocean surely remains a resource—of millions of tons of food, of oil, and perhaps of the development of wind energy—and at the same time is also a receptacle, absorbing our carbon dioxide, our unexploded warheads, our sewage. But now this generation also knows that it is, and always has been, a latent threat, a sleeping dragon. It cannot be denied that the South Asian tsunami powerfully, if momentarily, challenged global complacency about human control of the sea, or even human ability to predict what it will do, what it can do.[9] Over and over, both the media and survivors have ascribed agency to the sea, as evidenced in the accounts gathered by Waldman and Rohde and many others. The "fury" of the sea has become a commonplace—it is called the "angry" sea, the "hungry" sea, the "punishing" sea—which itself is now cursed. It is the sea that broke trust, that washed away "all evil"—whatever stood in the way of clean, ritually ordered life on land—as well as the very structures of that life itself. It is the sea that washed away thousands of children, the hope and the future. It is the sea whose powers of purification must be now either be reinterpreted in potentially apocalyptic terms or abandoned altogether as a therapeutic idea.

On December 26, 2004, millions of people were unwillingly wrenched from their ordinary lives by the waters and entered the world of marine myth, of the great history of the sea, with its surreal extremity and its tales of immersion, loss, and unpredictable salvation. As one who also lives by the sea with my children, I dedicate this book to them. Their experience has shown me that the "covenant" made with the sea by coastal communities was not one into which the sea, for its part, ever entered. Even though I love the sea, even though I greet it each morning with joy, I will never again in my lifetime see it in quite the same way. No one who has seen those parents clasping their drowned little ones ever will.

Acknowledgments

T his book was born from the heart, and not without pain. My
gratitude belongs first and foremost to my family. Thanks to
my husband Bruce, himself a scholar of religion and theology, who
supported me through the long hours, days, and months as I worked
on *The Sea Can Wash Away All Evils.* Bruce attended to our children,
Christina and Rosemary, with love and patience through a time
when their mother could not give them the attention they needed.
To Deanna Polizzotti, Rosemary's nanny, words cannot express our
gratitude for the devotion with which you cared for our little girl as
though she were your own while I worked long hours on this book.
Thanks to Susan Carter, who brought music, magic, and laughter
into my children's lives. My beloved daughters, I hope you will some-
day understand that, strange as it sounds now, I did this for you.

Words are inadequate to express my debt to Wendy Lochner, vi-
sionary and gracious senior executive editor in religion and philosophy
at Columbia University Press. Her editorial assistant, Christine Mort-
lock, also has been a great pleasure to work with. My sincere gratitude
belongs to copyeditor Margaret Yamashita and production editor Leslie Bialler. Leslie has been a heroic companion across rough waters.

I offer my heartfelt thanks to the scientists who constructively reviewed various sections of this book: in particular, James McCarthy, Agassiz Professor of Biological Oceanography at Harvard University; Christopher Leahy, Gerard A. Bertrand Chair of Natural History and Field Ornithology at the Massachusetts Audubon Society; marine acoustical scientist Michael Stocker of the Earth Island Institute; and the late Edward Dixon Stroup, Professor of Oceanography at the University of Hawaii. My thanks also to religion scholars Mary Evelyn Tucker, Rachel McDermott, Albert Henrichs, Jon Levenson, Peter Machinist, Elizabeth Pritchard, and Michael Puett, who read the manuscript and offered invaluable criticism; and to my students and former students who provided commentary and research support: Narelle Bouthillier, Ananda Rose Robinson, Rachel Smith, and Robyn Walsh. Without the insistent confidence of Robert Bosnak, Vijaya Nagarajan, Ronald Thiemann, Alexandra Kubler-Merrill, Sarah Coakley, Bernadette Brooten, Jack Hawley, Paul Waldau, Gretchen Hermes, and especially Michael Puett—Sam to my struggling Frodo—I would not have had the courage to begin this project, let alone finish it. My friends Linda Barnes, Erika Schluntz, Cecily Johnston, and Courtney Bickel Lamberth were faithful to the end. At the end, a thread lost for thirty years was strangely caught up again by the Great Weaver, when Thornton Davidson surfaced along the coast of Maine to bring me strength.

Thanks go to my students in the 2004 seminar I taught at Harvard Divinity School on the ocean in the religious imagination, "The Deep: Purity, Danger, and Metamorphosis." Your intellectual energy and spiritual depth were an inspiration. Special thanks to E. Garry Grundy III and David P. Charles from that class, who took the time to write extensive and invaluable critiques on an earlier version of this project.

Finally, I thank my treasured and gifted friend, the Reverend Carla Pryne, founder of Earth Ministry in Seattle, who has always loved the rich symbolism of water, has read my writings on the sea with care and enthusiasm, and has fought and prayed for the preservation of the earth, the seas, and all the life they sustain.

THE SEA CAN WASH AWAY ALL EVILS

CHAPTER I

The Dutch Bread-Man

Ocean as Divinity and Scapegoat

A controversy of quite modern as well as strangely anachronistic dimensions raged during the summer of 1992 in Holland. Its focus was a plan called the "National Gift to the Sea." The idea was to tow a 100-foot welded steel–framed human figure, with arms raised and stuffed with 20,000 loaves of bread, out into the North Sea and sink it there as a thank-offering. The bread-man was constructed on a giant dike north of Amsterdam and packed with locally baked loaves purchased by members of the public for $7 each. The half-million-dollar enterprise was sponsored by the Cargo Foundation, a group of ten publicity-shy Dutch people. According to organizer Kees Baker, the figure was "a sacrificial statue and an important expression of Holland's cultural identity. Holland has grown rich from the sea and a third of the land has been reclaimed from the sea. This is a grand gesture and a wonderful symbol."[1]

The Waterways Ministry, responsible for the purity of Dutch territorial waters, banned the project as violating the Seawater Contamination Act. The ban provoked 33,500 letters of protest. But equally condemnatory on the other side was Netherlands Greenpeace, whose spokeswoman Françoise Verdeuzdonk asserted, "It's just the most mindless,[2] primitive act imaginable. . . . It's pure pollution."[3] Baker's reply was telling: "They've just misunderstood the whole idea. This is an offering to the sea, in return for all we've taken from it over the years. It's a positive act, not a piece of senseless vandalism." If the ban was upheld, the proposed "solution" was to sink the carbohydrate-laden anthropomorph *outside* Dutch territorial waters.

The unreflective dumping into the sea of the waste generated by an industrialized global economy is not in the same category as the sinking of an 880-ton steel *orans* figure, an act lacking any secular veneer. There is a common attitude, however, reflected by both phenomena: the idea that the sea can always "take it." The ocean is beloved by coastal peoples like the Dutch, and yet it also is perceived as awesome, infinite, self-renewing, and never in need of protection itself. Or so it looks and seems to act, and thus so we think it is. Despite the environmental data, we who comprise human societies continue to behave as though we do not need to account for our actions toward the sea, partly because of the sea's paradoxical combination of familiarity and strange mystery, its literal "unfathomability."

Familiarity and Strangeness

The sea is everywhere, all around us, and many, even those who have never seen it, yearn for it in seemingly primordial ways. It is hard to escape: 74.35 percent of the earth's surface is covered by water; 70.8 percent of which is salty ocean containing 1,370 million cubic kilometers of water. The sea was the source from which our marine ancestors evolved millions of years ago "and yet carry still within us,"[4] dwelling as we do in salty amniotic fluid for nine months while our temporary fetal gill slits recapitulate phylogeny. Our lungs, filled with saline for weeks in utero and highly resistant to inflation, are the last of our organs to develop. Fish until we emerge, we can extract oxygen from air easily only at birth after a full-term pregnancy. The salinity of our blood is the same as that of the sea. Because saltwater is the fluid currency of our organism, we seem to remain marine creatures, "the salt which is in seawater is in our blood and tears and sweat."[5]

Perhaps this is the reason for our archaic sense of reunion. In 1921, after a long separation from the Atlantic Ocean, William Faulkner encountered it once more. He wrote to his mother,

> Then suddenly, you see it, a blue hill going up and up, beyond the borders of the world, to the salt colored sky, and white whirling necklaces of gulls, and, if you look long enough, a great vague ship, solemnly going somewhere.
>
> I cant express how it makes me feel to see it again, there is a feeling of the most utter relief, as if I could close my eyes, knowing that I had found again someone who loved me years and years ago.[6]

Faulkner captures the strong, oddly old sense of homecoming that can attend any return to the sea—in whose waters we cannot survive for more than a few hours, but which powerfully nonetheless exists as "a reservoir of private imagery and public myth."[7] The sea is the source of much of our metaphorical language, including religious metaphor, yet there is no metaphor for the sea. It is sui generis, comparable to nothing else, except perhaps remotely and only in certain respects, to outer space (see chapter 5, on the naming of the new "planet" Sedna). Humanly uninhabitable as a "place"—since we lose our gills at birth and no longer swim to move, eat, or mate—the sea remains our original home and matrix. From this "starting point," as she calls the sea, feminist biblical theologian Catherine Keller passionately advocates for a "tehomic theology" (from the Hebrew *tehom* of Genesis, "the deep" over whose face the darkness is spread). Such a theology, in Keller's view, affirms rather than rejects oceanic properties. For her, these are moral as well as aesthetic values, gathered up into a "true" description of the divine. Marine qualities become metaphysical dimensions as well as moral imperatives: watery chaos, void and abyss, indeterminacy, mutability, inchoate expression, "leakiness," nondifferentiation, teeming multiplicity, and above all, continual "becoming." This, Keller says, is our maternal experience of the divine, suppressed in patriarchal theologies that have sought to disarticulate religious experience through the countering control of hyperarticulation. This, she implies, is older: our forgotten marine birthright.[8] In what might almost be called the idea of geophysical nostalgia, Keller's *Face of the Deep* proposes to find the resources for radical theological revision in the lost sea.

In the image and meter of free verse, American poet Mary Oliver, whose home is Cape Cod, treats the same theme in her poem, "The Sea": humanity's source in the sea and the idea of our bone-deep, impossible nostalgia to return and live in it once more.

> Stroke by
>   stroke my
>   body remembers that life and cries for
>   the lost parts of itself—
> fins, gills
>   opening like flowers into
>   the flesh—my legs
>   want to lock and become
> one muscle, I swear I know
>   just what the blue-gray scales
>   shingling
>   the rest of me would
> feel like!
>   paradise! Sprawled
>   in that motherlap,
>   in that dreamhouse
> of salt and exercise,
>   what a spillage
>   of nostalgia pleads

from the very bones! how
they long to give up the long trek
   inland, the brittle
      beauty of understanding,
         and dive,
and simply
   become again a flaming body
      of blind feeling
         sleeking along
in the luminous roughage of the sea's body,
   vanished
      like victory inside that
         insucking genesis, that
roaring flamboyance, that
   perfect
      beginning and
         conclusion of our own.[9]

The longing for the sea, Oliver believes, is a longing for home, for
the mother of all life, and even for the fishy bodies of our ancestors.

But the sea is not home any longer. Despite our origins, it is chaotic
and lethal to human beings. It is a place whereupon we cannot build.
In many traditions, it was a place where one also could not properly
worship, could not be sure to have one's prayers heard or one's sacrifices accepted. The sea is unfamiliar—beyond us—and hence subject
to the abuse of "the other" seen in human social contexts. In his lyrical *A Sand County Almanac*, naturalist Aldo Leopold wrote, "We can
be ethical only in relation to something we can see, feel, understand,
love, or otherwise have faith in."[10] What then of the sea, of which we
can only see or understand a fraction? Environmental ethicist Clark
Wolf commented in response to Leopold's principle,

Most of us cannot regularly "see" or "feel" marine ecosystems. It
is certain that we will never fully "understand" them in their full

complexity. Surely it is easier for most of us to understand ourselves as "plain citizens" of terrestrial ecosystems. When we enter a marine environment—often importing with us a compressed version of our own non-marine environment—we may feel instead like alien foreign visitors. For these and other reasons, it is easier to generate public concern about deforestation in the Rockies. If we cannot, in Leopold's terms, "see, feel, understand, love, or otherwise have faith in" marine environments, does it follow . . . that we cannot be ethical with respect to these environments?[11]

Wolf offers the Kantian-based dichotomy of intrinsic and instrumentalist views of the natural world. In his "humanity imperative," Kant insisted that persons must always be treated as ends in themselves and never only as means. In other words, as bearers of rational nature, human beings have intrinsic value and cannot, with moral impunity, be treated instrumentally. However, as Wolf observes, "Kant rejected the notion that the natural world could have intrinsic value and would have rejected the claim that there is something morally objectionable about using natural systems as mere means."[12] Since, at least as far as we are aware, the sea does not possess rational nature, and thus, following Kant, cannot be ascribed any intrinsic value.

For marine coastal management scientist Adalberto Vallega, the view of the sea as a means to human ends strongly characterizes both modern and postmodern societies, though with important differences.

Modern society perceived the ocean from a very hedonistic perspective, as an enormous reservoir from which to supply human communities with food, energy and mineral resources. Post-modern society has acquired a broader perspective—the oceans as a planetary component blessed, not only by enormous resources and a hidden wealth of cultural heritage, but also marked by physical and ecological processes."[13]

Modernist views of the sea were sectoral, considering physical features and processes separately, whereas postmodern society sees the sea more holistically, as a "complex interactive system."

> The conventional, modern approach to the ocean was thus directed towards picturing how the ocean environment was physically constructed. This approach has been largely superseded by one aimed at representing how it works.[14]

Even as the gaze shifts from what the sea is made of to "how it works," we remain most interested in what we can get out of it or put back into it without consequences. Instrumentalism is the feature shared by the models of both periods, which have "tended to consider the ocean as a new planet within which human activities may expand, and from which it should be possible to recover the essential resources necessary to meet the demands of future generations."[15] Knowledge of how the sea "works" is accumulating simultaneously with growing indications that it cannot "work" in these same ways indefinitely.

Instrumentalism accurately describes long-standing secular practices of polluting the sea, but it also might partially exegete the role of the scapegoat in sacrificial systems. Scapegoats also have no "intrinsic" value, whereas gods do: gods are the ultimate source of agency and value in both monotheistic and pantheistic religious systems. Rather, scapegoats are useful and valuable only in relative terms, insofar as they are believed to be able to "draw away" from the group its pollutions, tensions, or other forms of anxiety-producing dis-ease. Classically, the group "restores" its imbalance by exiling and immolating the human or animal scapegoat—driving it into the desert, as in the Yom Kippur ritual of the Jerusalem Temple, or the sea, as in ancient Greek ritual texts from Colophon of the *pharmakós* or the synoptic accounts of the herd of swine who rushed into the sea of Galilee, bearing in their bodies the demons Jesus had cast out of the Gerasene demoniac (Mt 8:28–29:1; Mk 5: 1–20; Lk 8:26–39). Along these lines, in *The Love of Nature and the End of the World*, following René Girard and Gil Bailie, Shierry Weber Nicholson argues that elements of the environment have served as violent sacrifices, producing all of the culture-building benefits of "group cohesion achieved through scapegoating . . . order, stability, and the like."[16] The question arises: Over the centuries, has scapegoating the sea, by taking its cathartic powers for granted, enhanced "order" across cultures by allowing them to discard undesired products? Is what was once a central religious idea now also manifest in human patterns of consumption and disposal since the start of industrial age?[17]

In the peculiar yet revealing episode of the Dutch bread-man, the sea played two roles, depending on one's perspective, as a divinity (the recipient of a major sacrificial offering) and as a scapegoat. A spokesperson in the Netherlands Embassy in Washington told me in the fall of 1992 that it was not the issue of marine pollution but the popular outcry over the moral implications of bread "thrown away" in the face of world hunger that ultimately stopped the sacrifice. Equally compelling, however, to those who followed this story, was the implied vision of the sea itself as a bottomless receptacle. The bread-man story represents the collision and conflation of two concepts of "pollution," whose alleged conceptual and functional separability this book challenges.

Religious Pollution and Environmental Pollution

The first kind of pollution in the account of the Dutch bread-man is ancient and collective: the impulse to "pay back" the divinized ocean with a gift yet never questioning its ability to absorb the statue and its contents. Religious notions of purity and pollution, particularly for oceangoing or shoreline peoples, chronically regard the sea as the most powerful vehicle for catharsis, which can absolve—and dissolve—moral and ritual contamination, often undifferentiated in many traditional systems.[18] The ocean is the place that can make religious contamination literally disappear.

The title of this book is taken from the words of Iphigenia, the daughter of Agamemnon and unwilling high priestess of human sacrifice in Euripides' *Iphigenia in Tauris.* When ordered to kill her own brother Orestes, himself a matricide who lands "by the dark sea-wash" in Tauris, Iphigenia ceremonially sprinkles him with the water of the sea, thereby making him pure and acceptable. As she explains to the Taurian king, Thoas, who doubts his suitability as a sacrifice to Artemis: *thálassa klúzei pánta t'anthrópōn kaká*—"the sea can wash away all evils of humankind." The world's religions have a long inventory of ritual responses to the sea. Most, although not all, construct the sea as infinite and as supremely cathartic, diluting and carrying off what is ritually impure in human religious systems and thus dangerous to human well-being. Throwing things into the sea—the Greeks even had a verb for it, *katapontízein*—allowed quotidian social and religious life to continue on land. The sea thus enabled a collective terrestrial life and its attendant values.

Historically, the sea has received not only religious contamination. According to many sacrificial texts, what is "given" to the sea is more frequently not an ambiguous votive but the offal of human lives and enterprises. The second meaning of "pollution" is the one used in environmental discourse. Hence the second element of the case of the Dutch bread-man is the hallmark of a pressing concern in global ecology: the objection to dumping an "alien" and extraneous object into the ocean. Every dump of solid foreign objects or discharge of hazardous or biochemical waste alters the composition of the sea. Such practices threaten the oceans as life-supporting matrices, which in turn threaten the viability of the planet as an ecosphere. Marine pollution is rapidly becoming one of the greatest threats to the health of the planet, as the oceans and their swarming zootic populations are compromised, staggering with the weight of what they are asked to absorb, and poisoned by the toxicity.

The anthropological study of ritual pollution and contemporary scientific research in environmental pollution apparently share no premises, no central questions or methodologies, and no content.

This book contends, however, that in fact these two fields of inquiry converge in the sea. However culturally constructed, the drive to treat the sea as a receptacle for pollution is common to both the exigencies of religious purity and human habits of waste disposal. In other words, we might look beyond the apparent difference between *ritual pollution*, an esoteric subject that is the province of anthropologists and religionists, and *environmental pollution*, the province of scientific and political ecologists. I suggest that the natural properties of the ocean—and the resulting human illusions about what it can "always" do—have persistently inspired and aggravated both kinds of catharsis. What unites the ancient religious view of the ocean as the ultimate cleansing vehicle for ritual stain and the modern view of the ocean as a vast flushing (and self-cleaning) toilet, a receptacle for every kind of waste, is a fundamental, shared thought: because it seems infinite in size and eternal in life span, because it is made up of salt and water, because it is in constant motion through the tides, waves, and currents that carry things away, the sea will always dilute or change what is put into it, and not itself be changed.

In my view, it is insufficient to deplore such practices or to work to alter them only through national and international legislation. As in the case of all social habits—and all religious practices— a thorough exegesis of the inner logic at work is an important preliminary component to change. That logic, I believe, is based on the wide range of human perceptions of the physical properties of the ocean as a natural entity. Such "logic" or sets of assumptions have not been "handed down" through civilizations, nor do they exist in the collective unconscious. Albeit surely culturally determined, the idea of the sea's infinite capacity to dilute, cleanse, and neutralize is first naturally formed—and constantly reinforced—by human experience, however culturally determined, of the ocean itself. The sea continually "presents" itself to the human senses as an infinite cathartic agent, not as a passive recipient whose purity and biotic abundance can be compromised. It simply does not occur to us that there is anything the ocean cannot absorb or wash away.

Rather, the sea "seems" unassailable. Ocean is an autonomous, separate domain, a vast, moving body of water whose nature as "place" is so qualitatively different from anything on any land, the continents, the archipelagoes, the islands of ice or lava, that it apparently cannot be affected by human activity. One of the more interesting trajectories to have emerged from postmodern thought is the theorization of space itself, no longer understood as neutral but as socially "produced," culturally "made." In the rebuke of Ron Shields,

> Failing to examine the nature of space as a cultural "artifact," the realm of the spatial has often been assumed to be purely neutral and a-political, conferring neither disadvantage, nor benefit to any group. This "empirical space" is complacently understood to be fully defined by dimensional measurements . . . [and] excludes important cultural and cognitive issues from consideration. These issues direct our attention to the manner in which "space" is part of a culturally created system of philosophical categories. It is a "place-holder" when no objects are present and an index of totality of numerous co-present objects and elements in empirical-spatial relationships.[19]

The ocean has been a paradigmatic icon of space—endless, virtually unbroken space in contrast to the highly differentiated topography of land. On a globe, as James Hamilton-Paterson put it, the blue swathes of seas "have their own coherence as two-dimensional representations of *not-land.*"[20] Unstable and uninhabitable, the ocean will never be a place of "built" culture. Using the terms of Shields's analysis, it is thus a "place-holder" where no permanent objects of cultural significance are present. However, following Shields, the ocean is not neutral at all, but as not-land or no-place, it is instead a highly constructed "space" whose very nature is its alleged identity as a vacuum. As the antithesis of "land," or of "place," responsive to lunar and solar cycles as well as those of Earth, the sea seems to possess a fundamental autonomy, an oppositional identity that has been construed as anticultural and anarchical. This malignant autonomy is reflected in ancient religious narratives in which the sea was a volatile divinity in its own right. In Mesopotamia, the Levant, and Israel, all desert societies, it was a divinity that had to be subdued, contained, or even slain, so that cultural and cosmic order might be established.[21]

The sea thus offers the perfect means of catharsis. As an independent, allegedly noncultural nonplace, it can wash away all that impedes the human establishment and maintenance of civilization. This deeply held, experientially based idea about the sea can be found unexpectedly embedded in familiar places. Consider Lord Byron's famous marine paean from "Childe Harold's Pilgrimage":

Roll on, thou deep and dark blue Ocean—roll!
Ten thousand fleets sweep over thee in vain;
Man marks the earth with ruin—his control
Stops with the shore;—upon the watery plain
The wrecks are all thy deed, nor doth remain
A shadow of man's ravage . . .[22]

That "man's control" does not stop with the shore, and that his ravage is more than a "shadow" upon the watery plain is, in many ways, a relatively new situation, the result of the global economy's industrialization. The very fact of meaningful human impact on the sea, I would argue, represents not only a new reality but also a new idea, an apparent contradiction in terms, one which remains largely unassimilated into global consciousness. The notion of a finite or fragile ocean may register in social awareness on a rational level among environmentally educated élites, but perhaps only in those countries in which the exigencies of coping with civil war, genocide, or economic problems do not already consume policymakers. On a deeper level, and most certainly on a popular level, I believe that this idea remains counterintuitive. That is, until we acknowledge the cognitive dissonance between the data of environmental mandates, which are contemporary, and the natural experience and perception of the sea, which are primeval and remain largely unchanged until now, except on the days when the hypodermic needles wash up and close the beaches, we will be unable to halt the destruction of the earth's great salt bloodstream, which the Purāṇas personified as Samudra, "lord of all rivers."

My goal is to begin an exploration of human constructions, both sacred and secular, of the ocean's qualities. The idea of marine cleansing has been manifested in myriad ways throughout religious history, and I will not attempt an exhaustive catalogue of this linkage between sea and purification. Rather, I will present three very different cases of comparable religious responses to a natural element. I do so in the hope of demonstrating that there may be a phenomenological similarity between historically religious ideas about the sea as a limitless protective vehicle of ritual or moral catharsis and the global attitudes that contribute to marine pollution.

The ocean is, as Homer wrote, "full of fish." The idea of the sea as a vast laundromat or a great toilet seems to conflict radically with that of its role as a superabundant womb. The two ideas actually differ very little in certain respects. Like marine catharsis, marine fertility—although not my focus in this book—is an idea also often seen in religious metaphor and persisting to the present. Human responses to this symbolic field of generativity have resulted in the grave depletion of the sea through overfishing,[23] what natural sciences journalist Deborah Cramer calls "the myth of inexhaustibility."[24] In the latter half of the twentieth century, the annual global harvest of marine protein for human consumption from capture fisheries and aquaculture rose from 20 million to nearly 71 million metric tons, and a harvest of 106 million metric tons was estimated for 2005.[25] Depletion and extinction have become grave threats despite the initiatives in the past few decades to manage biotic resources. In the case of the twin marine myths of endless cathartic powers and biotic inexhaustibility, religious and environmental responses to the sea can be seen as comparable as well. Both are catalyzed by views informed by collective experience and empirical observation of coastal life, but no longer by the larger scientific reality: we are polluting the sea to death, on the one hand, and fishing its humanly consumable species to extinction, on the other.

Mircea Eliade and Mary Douglas

At the heart of this book lie certain specific strands in the thought of two monumental and highly contested figures in the study of religion, Mircea Eliade and Mary Douglas. Despite numerous critiques over the past decades, the work of both must be contended with to this day. While my argument has been catalyzed and informed by these thinkers, I also challenge and problematize their ideas. As a religious phenomenologist of natural objects, with his strong emphasis on patterns in religious response to universally perceived qualities of those natural objects, Eliade is central to my project, although I would disagree with his position that for "archaic man"—or for postmodern human beings—the direct experience of the cosmos as transcendent or powerful sequentially precedes its specific cultural construction. Instead I propose that these occur simultaneously in any human mind and are not epistemologically distinguishable. As an anthropologist whose work has provided our most important paradigm for the construction of purity and pollution in religious systems, Douglas is an equally inevitable conversant with anyone who wants to exegete these issues. My own analysis begins with Douglas's argument that "dirt" is a relative rather than an absolute term, defined by what "defiles" in a particular social system. Furthermore, as she maintains, the concept of "dirt" is always inextricable from, or even determined by, what is also morally "cordoned off" in a given culture. Where I diverge sharply from Douglas's structuralism, paralleling certain criticism of her work in British social and environmental theory, is in my rejection of her view that dirt (or pollution) is purely a socially constructed category, an arbitrary indicator of position on a graded hierarchy that lacks any corresponding objective reality. This is solipsism writ large, born in anthropological discourse but then detrimentally applied to environmental policy. In my view, a more realistic encounter with pollution as a danger to the physical viability of the planet and of the human race becomes crucial to the task of confronting marine pollution for what it is, as well as where it originates.

Eliade and the Religious Phenomenology of Natural Objects

In *Patterns in Comparative Religion,* Eliade proposed that in religious thought, natural phenomena are sacralized insofar as they manifest primordial power.[26] In his later *The Myth of the Eternal Return,* he wrote that religious archetypes had their genesis in natural events, sacralized through repetition.[27] Although Eliade and the wider field of phenomenology have been criticized in recent years for ambiguity regarding the assumption of an a priori category of the sacred, in the face of that criticism we can lose sight of the fact that Eliade high-lighted something inescapable in the history of religions. Certain features of crucial natural elements do lend themselves consistently to particular, often-repeated metaphysical patterns of interpretation.

There is a reason that the sun that disappears in darkness each night and miraculously reappears on the opposite horizon each morning was associated with myths of death, rebirth, and anxiety in shrines from Ise to Tenochtitlan to Heliopolis. Its burning disk in the sky, so much like an all-seeing eye, has made solar gods everywhere also divinities of justice, like Shamash and Helios. The Nez Perce had a saying at the start of each day: "It is morning! We are alive! The sun is witness to all that we will do today!" Solar gods often are gods of evidence, for the sun shines without prejudice and does not fail to notice anything, like the rape of a goddess's daughter by the god of death driving his black chariot. The sun remembers and redresses.

Each natural element has its own logic, various dimensions of which are observed and, depending on the cultural context, are conscripted into social and religious systematic thought. There is a reason why trees, whose form and size sometimes resemble the human body, have often been perceived as ensouled: why they bleed and weep in Manichaean texts of the third century C.E.,[28] why they march on human warriors in Celtic myth,[29] why they harbor human embryos among the Warramunga in Australia, why they self-defensively throw apples at Dorothy in *The Wizard of Oz* and why, during the last few decades, they have been clothed in saffron robes and ordained by the ecologically oriented Forest Monks of Thailand. We do not need to invoke a theory of the collective unconscious to explain this persistent human habit of seeing trees as people. We need only look at a tree and sit at its roots for a few seasons, watching. The natural characteristics of trees lend themselves to certain pathways of symbolic, culturally inscribed interpretation. But they also are naturally reinscribed by daily human interaction with the natural world.

While far from representing fundamental types that ultimately extinguish all differences among religious thought-worlds, such ubiquitous patterns nevertheless challenge certain features of Kantian epistemology—or at least the way that Kant's ideas about the limitations of human perceptions are deployed in neo-Kantian efforts hermetically to seal different religions' views of the same natural element. It may indeed be possible to speak of the same sea—or the same seven seas—as constructed in culturally specific, even unique, ways, yet as sharing a number of consistent features that can be described and considered. Here Eliadian method may be modified by a more empirically based methodology, one that starts with the manifestations of the natural element itself.

By no means do I intend this work as an intellectual genealogy. I believe that it is impossible to demonstrate a "history of ideas" that has evolved into the contemporary construction of the ocean. Instead, I would ask how natural elements have been cast in efficacious roles in religious practice and what the epistemological basis has been for those symbolic roles. From that basis, recognizing that epistemologies vary from culture to culture, we may begin to discern how a natural element is religiously constructed and the shape of its role in a particular world of thought. If that pattern persists across cultures, then we must account for it. We need not resort to archetypes, but it verges on blindness to dismiss "patterns in comparative religion" as generalizing, reductive overlays of colonialist value systems.

It is perverse to insist that the Pāli rabbit, the future Buddha, who leaps into the fire to save a starving Brahmin in the *Jātaka* tales and whose extracted image Sakka places on the face of the moon,[30] cannot be compared with the Chinese rabbit on the moon who pounds the elixir of immortality, or the Nahuatl rabbit who is flung by worried gods into the face of the overbearing Aztec moon, Tecuciztecatl.[31] Our overdeveloped preoccupation with contextual specificity may occlude the reality that when seen from the earth with one's head tipped to the side, the shadows of the moon's surface do in fact look like a great lunar rabbit, right down to the long ears. This is a monthly occurrence, visible for days around the world. That cultures without any historical contact have independently seen and mythologized a rabbit in the moon should be neither surprising nor contested as a move toward "universalizing." Rather, the pattern is explicable by looking at the moon.

Eliade recognized and asserted these complex correspondences, although he also found in history some kind of "pure" response to natural elements, an experience (of the sky, the sun, the moon, the mountain, the tree) that took place prior to the processes of religiously constructive imagination. Eliade theorized a "first response," an unmediated experience of the sacred. For example, as Allan Larson observes, Eliade wrote in *The Sacred and the Profane* of the sky's limitlessness (corresponding, for our purposes, to that of the sea) and argued that such celestial infinity "spontaneously becomes an attribute of divinity."[32]

> This is not arrived at by a logical, rational operation. The transcendental category of height, of the super-terrestrial, of the infinite, is revealed to the whole man, to his intelligence and his soul. It is a *total* awareness on man's part; beholding the sky, he simultaneously discovers the divine incommensurablity and his own situation in the cosmos. For the sky, *by its own mode of being*, reveals

transcendence, force, and eternity. It *exists absolutely* because it is high, infinite, eternal, and powerful.[33]

Problematically, and betraying the cognitive imperialism that Lévi-Strauss challenged by demonstrating the thought of "primitive" Brazilian peoples to be as self-reflexive and logical, within its own framework, as that of modern Europeans within theirs, Eliade wished to place this process beyond the operations of "reason." But more importantly for our purposes, he offered the possibility, "long ago and far away," of epistemological experience that can somehow translate into theological form without passing through the travail of conceptualized, which is, I would argue, a culturally mediated act. As Larson puts it,

> For the archaic man the experience of the sacred is found in the other, in the sky itself before any conceptualizations, before any flights of the imagination. . . . Eliade does not deny the significance of the social sciences, of history, but he continually tries to reach down beneath the presuppositions to the actual experiences upon which these sciences are based.[34]

Can there be any access to the "actual" experience of natural elements before they are conceptualized or theologized? In this respect, Eliade's is not the model I am proposing. Rather, I offer that while direct experience of the sea in its uniqueness is fundamental to both ancient religious and contemporary social categories of pollution and has constantly reinscribed both, such an experience has never meaningfully preceded its highly particularistic conceptualization. These operations are inextricable, like salt's presence in seawater.

Mary Douglas and the Social Dimensions of Ritual Pollution

In her ground-breaking book *Purity and Danger*, Mary Douglas wrote in 1966 that "ideas about separating, purifying, demarcating, and punishing transgression have as their main function to impose system on an inherently untidy experience."[35] Dirt is determined not by inherent qualities but by its role as "matter out of place," whose displacement away from the contaminated area and into the place of disposal is somehow permanent. "Reflection on dirt involves the relation of order to disorder, being to non-being, form to formlessness, life to death."[36] Extrapolating largely from Levitical codes but also from her own fieldwork among the Lele, Douglas argues that categories of "pure" and "impure" bear the intellectual weight of religious taboo. *Purity and Danger* rightly foregrounds the typical expression of social or moral concepts in the form of physical contamination: "pollution" therefore represents an idea of cultural disorder. Just as the sea has traditionally been a place of nonbeing, formlessness, and death, it has also been an appropriate "place" (or, as I have suggested, "nonplace") for pollution. I hope to show that in the past, as in the present, it was thought that the sea could dispose of and render harmless what was too morally, religiously, or physically dangerous to continue to exist on land. It is the ultimate "away" of the expression "taken away."

For a culture like that of the ancient Greeks, for example, as we will see in chapter 4, moral insult and ritual pollution were so linked as to be virtually inseparable. A catastrophe of hubris or wrongdoing could be inherited by future generations, just as a disease can. A curse with a moral genesis had startlingly physical properties, including communicability to subsequent generations of the family born long after the original offense. It thus required equally physical means of alleviation. We find such ideas irrational, and yet nearly every parent who has lost a child to death or jail for a heinous crime has stories to tell of the suddenly silent phone and the averted looks in the grocery store. "It was almost as though they might catch it," a bereaved mother once told me in reference to her many friends who dropped their contact with her husband and her after their youngest son at age twenty-one died in his sleep of undetected myocarditis. The Greeks just called all these terrible things, both voluntary and involuntary, "evils": κακά (Iphigenia's word). Κακά comprised death, disease, and transgressions; they were collectively represented by the stain or contagion called *miasma*. Contact, however inadvertent, with someone afflicted by *miasma* could indeed infect another person, no matter how innocent of involvement with the original evils. The Greeks made no bones about this.

Douglas radically and rightly insisted on the embeddedness of concepts of pollution and purification in "symbolic systems":

> Defilement is never an isolated event. It cannot occur except in view of a systematic ordering of ideas. Hence any piecemeal interpretation of pollution rules of another culture is bound to fail. For the only way in which pollution ideas make sense is in reference to a total structure of thought whose key-stone, boundaries, margins and internal lines are held in relation by rituals of separation.[37]

She thus anticipated the more recent holistic view of the sea itself as a "self-organizing system" rather than as the sum of its parts and processes, as discussed in chapter 2. Starting in the early 1980s, not surprisingly, Mary Douglas felt compelled to respond directly to the application of her work to the environmental movement. At a lecture entitled "*Purity and Danger* Revisited," delivered in 1980 at the Institution of Education at the University of London, she described how the tables have turned. Natural elements have progressed from a state of moral authority in human social and religious spheres to passivity: "The earth's girdling waters and envelope of atmosphere are no longer the source of divine vengeance, visiting thunderbolts on liars, and floods and flames on godlessness. Without moral agency of their own, they are becoming the passive, vulnerable condition for life on this planet."[38]

Close examination of Douglas's rhetoric here, however, which she expanded a few years later in *Risk and Culture* (cowritten with Aaron Wildavsky in 1983),[39] reveals that she does not sponsor any shift away from the "constructed" nature of pollution, even when the arena of inquiry shifts from human cultures to the natural environment.

Hence she claims, "With us, no more than with our forebears, nature and purity are not technical terms: when the border uses them, the centre is being arraigned for causing pollution. When the centre uses them, a contagious border is being cordoned off."[40]

In *Risk and Culture*, Douglas takes the final step of relativizing industrial pollution by making it an "idea," one of the many socially produced "risks" attributed to technology and its use by governments. She compares the risk of environmental pollution, which she calls a "theory about the world," one that "politicizes nature," with the arbitrarily determined dangers of ritual pollution given in a particular religious system on the grounds that neither have objective correlatives. "The notion of risk is an extraordinarily constructed idea, essentially decontextualized and desocialized."[41] Here perhaps she goes too far, extrapolating from *Purity and Danger* in a direction that seems, in light of the evidence, fantastic. The current "risk" and threat *to* the sea and thus *by* the sea to long-term human viability is more than socially produced theory.

In sum, I endorse Douglas's revelation of the strong interconnections, even identifications, between religious binaries of purity/pollution and societies' coded moral hierarchies. But I question her thorough-going belief in the "socially constructed" nature of both ritual and environmental forms of pollution. Neither lacks hefty anchors in data-rich reality.

With these reservations, then, certain tools offered by Eliade and by Douglas, the wide gap between them notwithstanding, may help interpret the long human history of relationship to the sea.

Environmental Theology: An Uneasy Marriage?

Our distinction in English between the differing semantic fields of the two kinds of "pollution" just discussed has only recently been bridged, in one area by particular monotheistic "environmental theologies." After decades of silence in the face of a mounting ecological catastrophe, progressive sacramental Christian traditions are now the main source of such thought and rhetoric, but echoes can also be heard in their Jewish and Islamic counterparts. Environmental theology seeks to recast the material pollution of the planet as *sinful,* an affront to the Creator, because violently wrought upon the Creation. For example, in late 1998, Patriarch Bartholomew, spiritual leader of the theologically conservative Eastern Orthodox Church, expounded on a category of sins with roots in patristic theology but undeveloped in Orthodox theology until that point: sins against the environment.

> For humans to cause species to become extinct and to destroy the biological diversity of God's creation; to degrade the integrity of Earth by causing changes in its climate, by stripping the Earth of its natural forests or destroying its wetlands, for humans to contaminate the Earth's waters, its land, its air and its life, with poisonous substances—those are sins.[42]

Such theologizing may be seen as a self-conscious resurrection of ancient concepts that married moral stain and physical contagion, conceptualizing them as essentially one entity. The long divorce of religious pollution from modern marine pollution—the latter hitherto the province of science, environmental law, and "green" activism—renders the reunion of the two kinds of pollution under the traditional category of "sin" less than organic. Yet both are human activities manifesting the tendencies that Douglas identified, namely, the anxiety over what is disordered and "wrong" in human existence and the desire to dispose of it: the impulse to separate and purify, to move what is called dirt, "matter out of place," away from the inhabited space into the place of disposal. Both kinds of thought about pollution, I hope to show, have looked to the sea as a major source of catharsis. What greater line of geophysical demarcation exists than the line that separates the land from the sea, so what more likely candidate for these kinds of "necessary" separation and disposal?

Collective perceptions of the salt sea and the readiness with which it seems to "wash away all evils" have led to its destruction. As in the slow redress of hierarchical constructions of race or of the "acceptable" abuse of children, remedy can thus only lie in a reorientation of collective ideas about the ocean. It is in a substructural shift that is not only psychological but also epistemological, girded by religious notions of *metanoia* and then legally enacted and politically realized. Just as children are no longer the private property of their parents, although once they were, throughout the world, and so for all practical purposes remain in many cultures, the sea can no longer be our scapegoat. In support of Bartholomew's mission to the Black Sea, Metropolitan John of Pergamon, his chief theologian, argued at the end of the voyage of the *Venezilos* over the dying Black Sea, "The ultimate problems reside in the human heart and mind. The changes needed will only happen if human attitudes change. And what more powerful force for changing attitudes can there be than spirituality and religion?"[43]

Overview of Contents

In chapter 2 I summarize some of natural cathartic and "self-purifying" features of the world's oceans and then broadly outline the known facts about the anthropogenic factors from which they now suffer, including the role of ideological or conceptual factors. In chapter 3, I discuss the broader comparative context in which the sea has been seen as a purifier in the history of religions. Assuming the relevance of a phenomenological approach to the questions raised here, this chapter is "mapped" by examining the attributes of the sea that are open to a cathartic interpretation.

A closer examination of three traditions follows, where the salvific ocean to religious purification. Chapter 4 looks at the ritual traditions of ancient Greece dealing with the marine purification of cult statues and the absolution of the stain of blood guilt. Chapter 5 explores the Inuit complex of marine myth and shamanic ceremonial whose centerpiece is Sedna, *inua* of the seafloor, her hair made filthy by accumulated human transgressions. In chapter 6, in another religious world and on a different register of the idea of "purification," I turn to the apocalyptic accounts of the medieval Hindu Puraṇ as of the lethal fire released by the god Śiva and its absorption by merciful Ocean. The Puraṇic myth-complex of the submarine mare (*vaḍavā*) expresses the sea's unequaled ability to absorb harmful, world-destroying substances with apocalyptic potential.

Finally, in chapter 7, I argue that aspects of these traditions illumine, perhaps even ominously, the contemporary problem of marine pollution in unexpected ways. This is especially possible in the mythemes of the sea's revelation of what was once hidden in it or its rejection of what was discarded. These are the myths of "unwanted return," of the ocean's refusal of its assigned role. Kevin Hetherington, a Lancaster sociologist who has cogently criticised Mary Douglas for her view of dirt as purely representational, also takes her and Victor Turner to task for assuming that pollution somehow "really goes away." Hetherington calls upon Marilyn Strathern's understatement that "pollution surprises by its untoward nature, an unlooked for return; yet those involved in the activity of waste disposal know that one cannot dispose of waste, only convert it to something else within its own life."[44] Chapter 7 brings together religious and literary accounts of this reversal: the theme of the shocking nonpermanence of disposal into the sea, with almost always disastrous results.

The sea has always seemed able to absorb and neutralize poisonous substances. Now this *míasma*, which once had ritual or moral meaning or both and could sometimes threaten apocalypse, is constituted by something quite different: the toxic effluvia, including airborne contaminants that are the by-products of economic activity and development. But the need is the same: the washing away of what hinders normal life. The waste that we might never discharge into a stream or a lake—because we can see its opposite shore and therefore know that it has a finite capacity to absorb—is often pumped into the sea without check "as though to a bottomless pit,"[45] because we cannot see its other shore and do not believe that it has a bottom. Because of its tidal cyclicity, its great currents and hammering waves, we seem to believe that the sea must always wash itself clean. As we will see in the next chapter, this belief has begun to take a disastrous toll. As environmental journalist Michael Specter wrote presciently in 1992,

> With an immensity so humbling and so grand, few people can comprehend the kind of human assault required to do serious damage to the world's seas . . . [for] most Americans, along with their deeply held love of the sea has come an understandable belief in its intangible abundance. There may be a hole in the ozone and acid rain falling on the forests, but surely the oceans, the ineffably endless oceans—will always be there to wash away the accumulated grime of lives tethered to the land.[46]

This book is about that "deeply held love," along with that "understandable belief" gradually being exposed as a grand illusion by the threatened oceans themselves.

CHAPTER 2

The Crisis of Modern Marine Pollution

Once thought to be so vast and resilient that no level of human insult could damage them, the oceans are now crying out for attention.
Nicholas Lenssen, "The Ocean Blues"

Triumphant newsreels that played in theaters at the end of World War II showed aircraft carriers as they made their way home from the Pacific Theater, dumping their ordnance directly into the deep ocean. Just as one might conduct a marine burial, Navy personnel slid tons of explosives along planks and into the ocean. "The boys are coming home!" ran the voice-overs. "And they won't be needing their weapons anymore!" For complex symbolic reasons only partially accounted for by military ideology, the ordnance could not be returned to American soil. The war was over. The ocean, again "not-land," "no-place," provided a vast "safe harbor" for the transition from global conflict to American-led peace, absorbing the means of mass destruction that were banned from returning to their place of manufacture, thus re-creating the terrestrial United States as a nation beyond war. In 1994, older ICBM missiles, which require a target and cannot be safely dismantled, once aimed by the United States and the Soviet Union at each other, were redirected toward points in the Pacific Ocean. Should those missiles be accidentally launched, the sea will now receive and absorb their nuclear potentials. But

American and Russian societies will be safe, half a world away from each other on opposing continents. Here are echoes of the chorus in *Oedipus the King,* crying out to drive the war god into the sea, "to the great palace of Amphitrite or where the waves of the Thracian sea deny the stranger safe anchorage (196–97)."

What are the assumptions behind ocean dumping? How do we "envision" the fragile sea? Why do we continue to assume that it can absorb any danger, carry off any blight, dilute any poison, when the evidence to the contrary mounts steadily?

The mortal vision of the sea as an immortal means of catharsis predates the era of environmental crisis, manifesting itself in a deeper and older level of religious thinking. For the same reasons that human societies now pollute it, ocean water—and the ocean itself—were believed in many ancient religious traditions to have cathartic powers. The sea was and remains the purifier, "vast and resilient," as Lenssen labels it, apparently impervious. The definition of "purification" has changed from a ritual to a material one, although both may be certainly described as culturally determined; the ocean's ability to ceaselessly purify has also radically changed. Not only has the volume of waste changed, but also the type of waste, which affects the length of time waste remains in the marine environment.[1] That is the great difference between our times and the world of antiquity. The maverick archaeo-explorer Thor Heyerdahl describes a moment of harsh revelation during his journey across the Atlantic on a reconstructed ancient Egyptian reed boat in the late 1960s:

> Next day we were sailing in slack winds through an ocean where the clear water on the surface was full of drifting black lumps of asphalt, seemingly never-ending. . . . The Atlantic was no longer blue but grey-green and opaque, covered with clots of oil ranging from pinhead size to the dimensions of the average sandwich. Plastic bottles floated among the waste. We might have been in a squalid city port. . . . It became clear to all of us that mankind really was in the

process of polluting its most vital well-spring, our planet's indispensable filtration plant, the ocean.[2]

The "filtration" of which Heyerdahl spoke is not just a metaphor. Not only do human societies pollute the seas; the seas themselves, through their essential qualities, seem to be eminently pollutable because they are endlessly self-renewing. Oceans are natural circulatory systems like the human body's blood and lymphatic systems, distributing heat; recycling water into the atmosphere; breaking up, carrying away, and sinking foreign objects in their depths, and absorbing harmful toxins into their saline waters. As Dorrik Stow remarks, "organic substances, such as sewage sludge, can be dealt with by natural processes of bio-degradation . . . provided that the environment is not overloaded."[3] The same is true of thermal pollution and oil spills. But the marine environment is increasingly overloaded, and impacts often cannot be mitigated. "Like a living web, moving and winding and mixing, wrapping itself perpetually around the whole world,"[4] The currents constitute part of the actual ability of the worldwide sea to dispose of what runs into it, what is thrown into it. Soils and other coastal runoff, the heavier effluvia of rivers, are forced by marine currents toward the seafloor, where human, animal, and plant remains are carried as well, away from the surface.[5] In the meantime the hydrologic cycle, operating vertically in counterpoint to the currents' slow horizontal dance, draws up water from the salty sea, returning it to the earth in the form of 113,000 billion cubic meters of precipitation, two-thirds of which is returned to the atmosphere as it evaporates.

The world's oceans absorb great quantities of heat and release it slowly, thus stabilizing the earth's temperatures and, in conjunction with other factors, thermodynamically determining its various climates and weather patterns.[6] David Helvarg observes, "The top 2 feet of seawater contain as much heat as the entire atmosphere."[7] Through their constant cycling motion, the seas also distribute heat across the continental masses. The vast Kuroshio Current in the

Pacific Ocean transports heat from Japan to North America; the Gulf Stream moves north from the Gulf of Mexico and divides at the British Isles, contributing to warmer climates in northern Scotland and moving past Portugal to join the Northern Equatorial Current circling between South America and Africa. Massive macrosystems of oceanic currents, called gyres, are of circumferences so wide that they can redistribute heat from the earth's middle latitudes in the direction of the poles.[8] The South Pacific Gyre alone takes three years to complete a rotation. Yet as Heyerdahl warned, there are new limits. The Pacific Gyre has become contaminated with particulate plastic, moving slowly in a great whirlpool, a thick soup that, because of its momentum, cannot be diluted. It is unknown what effect this atomized plastic will have on the fish and other forms of sea life that ingest it.[9]

Equatorial heat, corpses, and minerals from the rivers are not all that the seas absorb. They have a surprising capacity that has been environmentally salvific and, until now, unrecognized. It was recently established that over the past two centuries the oceans have soaked up roughly half the carbon dioxide produced by human industrial activity.[10] Much of this has been assimilated by microscopic phytoplankton, one-cell plants dwelling in the sunlit upper ocean.[11] Over the history of life on earth, the photosynthesis of these oceanic plants has produced most of the oxygen in our atmosphere, which is essential to the survival of human beings and all other animals.

The biota of the seas far surpass those of the rain forests, which are sometimes referred to as the "lungs" of the world, as they sponge up greenhouse gases and release oxygen, thereby creating an atmosphere far more viable than it otherwise would have been. The results of a ten-year study were reported in two scientific articles published in July 2004. Besides showing that the oceans have absorbed about half the carbon dioxide produced by human activity, the first article also established that the oceans are moving the CO_2 into the deepest ocean, continually freeing up more absorbent surface area. In the words of geophysicist and coauthor Christopher Sabine of the Federal Pacific Marine Environmental Laboratory, "The oceans have a capacity to continue to take up CO_2 for thousands of years,"[12] although the remaining half continues to affect global warming, the melting of polar ice caps and the destruction of polar habitats, the die-offs of coral reefs, and the lethal rise in sea levels.

The second article indicates that the marine influx of carbon dioxide is actually changing the ocean's chemistry "on a global scale."[13] Natural acids are being produced in greater quantities, and the pH of the seas is falling, perhaps as much as four-tenths of a point by the end of the twenty-first century. Coauthor Joanie Kleypas, at the National Center for Atmospheric Research, remarks, "This kind of pH in the oceans that we're approaching is something the oceans probably haven't seen for 20 million years."[14] This drop in turn dramatically affects the rate at which organisms like corals secrete calcium carbonate skeletons. In laboratory experiments, the skeletons of such creatures become fragile or even begin to dissolve around the animals' organs. Kleypas estimates that the growth of corals will be stunted by 10 to 40 percent. Because acidity levels also affect coccolithophorids, a variety of marine algae, and pteropods, tiny shelled marine animals at the base of the marine food web, the impact may be far greater.

The ocean is not a place of silence. It is a vast acoustical environment—a world of myriad sounds produced by sea creatures, sounds whose human acknowledgment dates at least as far back as the archaic Greek tale of Arion, who was flung overboard by brigands but rescued from the sea by the dolphins when they recognized a fellow musician. Singing, clicking, whistling dolphins are joined by bassooning, fluting whales as the large, breathing, intelligent marine subjects of intensive contemporary investigation into the communicative values of noises.

Both the production of marine noise and the means of its perception by other forms of marine life are now being investigated by new technologies, largely pioneered by the military, as acoustical naturalist Michael Stocker notes. "The background noises that we took for granted as some indication of marine life are increasingly being re-valuated as the necessary sounds of animal survival—sounds that sea creatures use to communicate, navigate, hunt, bond, and breed."[15] In other words, the biological sounds of the sea may be highly adaptive and crucial to the perpetuation of many species, deliberately rather than incidentally produced. Not only do dolphins and whales sing, but fish grunt and scrape; from miles away, crustaceans and barnacles may be able to sense the sounds of their own species, opening and closing their appendages. We are just beginning to learn what these sounds mean and how they are "heard." One wide-ranging theory, called *acoustic illumination*, argues that "objects and features in water cast acoustic shadows and reflections of ambient noise that fish can perceive and integrate into the perception of their surroundings," just as sunlight allows us to see our surroundings by lighting them.[16] In many marine regions, these ambient sounds are now grossly interfered with, by "anthropogenic noise": especially sonar surveillance but also industrial noise, boating and shipping engines, and the underwater testing of explosives. As with every other form of human intrusion into its waters, we assume that the sea can absorb our noise without effect. But like our other assumptions about the sea's resilience, this one is turning out to be false. Many marine biologists wonder, for example, whether some mass whale disorientation and beachings may be caused by sonar interference with their navigational communicative noises.

The invention of the flush toilet in the late nineteenth century ushered in what seemed to be entirely a new concept: washing away waste into the waters. In environmental science this is known as "the victory of the flush." Until then, as is the case even today in places like rural China, night soil was usually returned to the earth. Now, however, the ocean has in a very real sense become the Great Toilet.[17] This "flushing" activity is most visible coastally, but its effects are most felt in the open and deep-sea environments. "Marine pollution" is a relatively recent concept, one that corresponds to, and yet lags behind, the escalation of the problem. It is only since the mid-twentieth century, with the scientific concern in the

United States, Britain, and the Soviet Union about the release of radionuclides into the atmosphere and the ocean that any significant awareness has existed. People have become aware that "renewable resources of the ocean could be placed in jeopardy as a result of man's activities."[18] It was only in the 1950s with the dramatic poisoning of the fishing population living around Minamata Bay in western Japan that impact of coastal marine pollution on public health became clear. For years the Chisso Chemical Company discharged methyl mercury chloride into the bay, where it was ingested by fish and shellfish and in turn ingested by families and their pets, often lethally ("dancing cats" in the streets being one of the first indicators of its severe neurological symptoms). Because Minamata Bay was an enclosed, protected marine ecosystem, it offered a flow chart of the ordinarily highly diffuse processes of marine pollution.[19]

According to the independent Pew Oceans Commission report, one of two major investigative panels issuing reports on the state of the world's oceans in 2004, "Paved surfaces have created expressways for oil, grease, and toxic pollutants into coastal waters. Every eight months, nearly 11 million gallons of oil run off our streets and driveways into our waters—the equivalent of the Exxon Valdez oil spill."[20] Polychlorinated biphenyls (PCBs) and mercury end up in the sea not only from rivers but also from atmospheric deposition, sometimes from tremendous distances. These artificial, carcinogenic compounds enter the human food chain when they are consumed by shellfish and finfish, and are estimated by the National Cancer Institute to be present in the tissues of 99 percent of Americans.[21] In just one year, 2003, fecal bacteria accounted for 18,000 days of beach closings and advisories across the United States. Each year, coastal development has caused the destruction of more than 20,000 acres of sensitive wetlands and estuaries across the country. The final report of the U.S. Commission on Ocean Policy, delivered to Congress in September 2004, noted that more than 110 million acres have been lost since the time of the Pilgrims.[22]

The degradation of more than 60 percent of coastal rivers and bays by nutrient runoff, caused when human development near shores and its attendant deforestation spill products from septic tanks, fertilizers, pesticides, pet wastes, viruses, nitrates, and phosphates into waterways, has resulted in widespread coastal eutrophication. At the sedimentary interface of estuarine and riverine systems, algae proliferate in response to nutrient enrichment. The algae produce more than ten to twenty times more oxygen through photosynthesis in sunlight than they consume.[23] These proliferating bacteria in turn consume the superabundant dead algal material, depleting the oxygen in the ocean water and killing both finfish and shellfish. The problem is especially aggravated in warmer waters, where less gas can be held in solution. As a result, "non-native species are increasingly invading marine ecosystems."[24] On the beaches at Malibu and elsewhere in Southern California, torrential rains like those at the beginning of 2005 and the resultant flooding often hasten the eutrophication by sewage-fed bacteria, as the California shoreline's outflow channels were built in the 1950s to empty waste and toxic agricultural runoff directly into coastal waters. Nutrient pollution creates a dead zone "the size of Massachusetts" each summer in the Gulf of Mexico.[25] Referring to the seaweed-choked lagoons of Venice, marine biologist Ivan Valiela remarked nearly thirty years ago, "Eutrophication is the single most important process taking place on the shorelines of the world, and it's happening everywhere."[26] Since that time, eutrophication in coastal waters around the world has only increased.

Neither are the open oceans immune. Along the shipping lanes of the high seas, an estimated 4 million metric tons of oil are spilled annually.[27] More than 2.1 million tons of liquid chemical waste are poured each year into the North Sea alone.[28] Even the vast oceans off Asia are now dangerously polluted with petroleum and plastic products, sewage, and urban waste, as well as radioactive contamination from continued nuclear weapons testing.[29] In the Arctic Ocean, near the archipelago of Novaya Zemlya, Russia has been dumping radioactive waste into the waters for decades, including canisters from nuclear submarines, creating what has been called "a marine Chernobyl" in the northern Pacific.[30]

What Vallega calls the "complex, interactive system" that is the sea is increasingly regarded as a single entity; a glance at a globe will show why. The "oceans of the world" are not plural; they are one. The oceans' interpenetration and chronic current movements make distinctions beyond a certain point meaningless. Accordingly, much scientific literature uses the term *World Ocean*. As Russian marine ecologist Aleksandr Souvorov finds, this single entity plays a major role in other global processes, and it is often the final player.

> The World Ocean is a "closing element" in all kinds of large-scale processes of substance circulation and transformation, whereas the oceanic branch of the biochemical cycles of vital elements plays an important role in the existence of all living organisms on our planet. That is why the protection of the ocean ecosystem, subject to appreciable anthropogenic impact which has already resulted in serious adverse environmental consequences, requires the joining of efforts from all maritime countries.[31]

As the "closing element" in such crucial processes, the sea is also the ultimate goal of equally harmful processes in its role as absorptive *pharmakós*, scapegoat, and receptacle of anthropogenic contamination. At the shore, where the land stops the rolling waves, it seems as though one can go no farther. But where air-breathing, uninsulated human beings cannot go for long without raft or diving gear, nuclear and biological waste, carbon dioxide, soluble and solid toxins, and noise can and do go farther, penetrating the sea's very heart and staying there. Thus the World Ocean becomes the unhappy end point of social and industrial as well as natural processes. Returning to the point made earlier in this chapter about the seas' apparent self-renewing capabilities, the fact is that *up to a certain point*, the seas can indeed self-purify. The problem is that we do not know what that point is.[32]

This was brought home in 1984 by the stark conclusion of a series of meetings entitled "The Workshop on Land, Sea, and Air Options for the Disposal of Industrial and Domestic Wastes," convened under the auspices of the Commission on Physical Sciences, Mathematics, and Applications of the National Research Council by Edward Goldberg of the Scripps Institution of Oceanography and Stanley Auerbach of the Oak Ridge National Laboratory. Among the meetings' goals was the assessment of the "assimilative capacity" of the world's oceans vis-à-vis foreign substances, including the impact of introducing into ocean waters any amount of a wide range of organic and inorganic elements. That is, what levels of a given pollutant (from municipal sewage sludge to hydrocarbons to lead, cadmium, or mercury) would be "safe" for marine populations and ecosystems and could be discharged into the ocean without killing or impairing life processes? Could predictive models be generated that would provide an index for irremediable damage? How much could the sea absorb before a price would be paid? As biological oceanographer James McCarthy, a participant on the workshop's Board on Ocean Science and Policy, comments, "Although many were skeptical of these efforts, others felt that the product could actually be a useful policy instrument."[33] Several years of intensive analysis of data through a variety of proposed macromodels produced a startling result. As McCarthy remembers, "It was an impossible task." In other words, the upshot of the workshops' conclusion was that *no* safe level for marine pollution of any kind could be determined. The concept of the ocean's "assimilative capacity" was meaningless.

The massive, dynamic entity that constitutes and even creates so much of our global environment is, because of its size, deeply "conservative" of its own nature—its thermal properties, its self-purifying abilities, its checks and balances and interrelationships. But like a great animal or tree that is hard to sicken or destabilize, once compromised, the complex sea is even harder to heal. Although "the World

Ocean's stability is high due to the tremendous volume of its water
. . . its balance is very difficult to restore once disturbed."[34] Biologist
John Cairns echoes the warning: "It appears highly probable that the
vast oceanic systems are quite fragile . . . and are protected primar-
ily through their vastness and the resultant dilution of all potentially
deleterious materials. Should an entire ocean be damaged, the time
required for recovery staggers the imagination."[35] Oscillating between
its natural role as "closing element" and the role into which human
activity has forced it, the ocean approaches end game.

CHAPTER 3

The Purifying Sea in the Religious Imagination

Supernatural Aspects of Natural Elements

I n *Natural Symbols,* Mary Douglas asserts that "the social experience of disorder is expressed by powerfully efficacious symbols of impurity and danger."[1] Perhaps no symbolic dichotomy better expresses this principle than the relationship between land and sea. Furthermore, this has always been the case; only the "disorder" has changed, from the destabilizing impurity of ritual transgression to that of chemical, biological, and domestic waste products. The anxiety, however, has remained. Disorder, as expressed by ideas of impurity, is too dangerous to tolerate or integrate in human social experience (although, *pace* Douglas, as I argue in chapter 1, impurity itself is real and is not simply a socially constructed vehicle for the experience of disorder). Terrestrial life can be maintained only by expelling disorder and impurity, thus reaffirming order and reestablishing purity. It has always seemed to coastal-dwelling human beings (and human communities have largely bordered on bodies of water; today more than 3 billion people live in close proximity to the sea) that the best place to expel the danger forever is the ocean.

I suggested in chapter 1 that the modern crisis of marine pollution, in all its variety and in all its range of motivations, has its source in age-old and persistent human reactions to the ocean's distinct and almost universally noticed attributes, some of which I described in chapter 2. The ocean is a body of water like no other. It is of incomparable size; it has tidal flows and currents; and its waves create a continuous, salutary motion so that it is never still but is always washing, heaving, eddying, sucking, and carrying away what it finds or what is thrown into it. It is a noisy, vast cauldron. It has always seemed like a living thing, with agency, intentionality, and moods—most intensely, in the thrall of marine disasters like the 2004 South Asian tsunami. The sea has always been assumed to be able to neutralize poisonous substances in any quantity because it dwarfed and drowned them; thus their danger was immediately diluted. That *miasma*, which, as we shall see in the following chapters, once had ritual meaning or apocalyptic import, is now constituted by something quite different: the unwanted waste of industrial production and the social existence of developed nations. In a sense, nothing has changed, because human perceptions of the sea, metaphysically translated and evident in religious texts and rituals, have remained constant, unexamined, and unaffected by a growing body of ecological evidence of the damage the sea is sustaining. These perceptions have now been "translated" into discharge.

At the deepest structural level, the level of thought "about" the sea and of what it is capable, traditional religious systems of purity, and modern industrial societies share the same need and requirement. This need is for the "permanent" disposal—out of sight, out of mind, and out of contact—of what land-based human societies in all their dimensions must be rid in order to maintain a normal, "pure" collective existence. Even in a largely secular society, the sea seems to offer the promise of ultimate catharsis and the related idea of personal and collective renewal. Both industrially and agriculturally based economies have been quick to exploit this promise, because human thinking about the sea is habitual. Modern economies are by no means explicitly informed by the beliefs of the ancient Greeks, circumpolar Inuits, or medieval Hindus; rather, they reflect analogous pathways of thought—comparable internal logics.[2]

The study of the history of religious responses to natural elements has its own history, one that in turn encodes deep tendencies and prejudices, especially in a field highly influenced by monotheistic ideas of what constitutes complex religious thought. Developmentalist ideas still structure our understanding of nature and religion, obscuring the intellectual content of traditions in which natural elements and forces played, or still play, a dominant role in systematic existential thought. "Nature worship" and "fear of the natural elements beyond man's control" are phrases characteristic of nineteenth and much early-twentieth-century scholarship on "primitive" or even "pagan" religions, no matter how theologically differentiated in reality. As we have seen, neither Eliade nor Douglas is completely immune, despite their sophisticated versions of these descriptions. Thus, as I have discussed elsewhere,[3] pantheons with theriomorphic gods have been assumed to be less conceptually sophisticated than those with anthropomorphic ones, and anthropomorphic gods in turn are evaluated as theologically cruder than abstract, ineffable ones. This interplay among animal–human–god in divine representations is part of a larger narrative whose goal is to show a "development" from the divinization—and propitiation—of the natural world.

This is not a new tension, but is embedded in the dialectics of many traditions. Nor is it untrue that some traditions have indeed consciously moved along the exact trajectory anticipated by this school of thought, with the natural world either fading in importance or undergoing conspicuous sublimation and subordination to "higher," more abstract powers, unanchored in the actual sun, moon, mountain, or tree. Particularly in the creation account but also throughout its exegesis and poetic representations, the Hebrew Bible adheres to a view of nature as created and perfectly controlled by the God of Israel, in pointed contradistinction to the imputed idolatry of its predecessors and neighbors with their storm deities and arboreal goddesses of fertility. Underlying this theology, as Jon Levenson has shown, lies an older, chronic tension observable in some of the biblical texts, in which the battle between God and chaos—and particularly the chaotic sea or sea monster—is far from over but must be refought periodically.[4] The subduing of nature by divine order is thus a biblical ideal, one not fully realized, even in biblical history. When the subject comes up, the New Testament also shows Jesus as the master of nature in the episode of the storm on the Sea of Galilee: "What sort of man is this, that the wind and the sea obey him?" (Mt 8:27).

By the classical period, ancient Greek religion saw the archaic winged Athena on vase paintings lose her wings to her familiar, the little Attic owl, and to the hypostasized Nike. Over the same period, Helios, the Greek sun god of clear Indo-European ancestry (compare the Vedic Sūrya) with a crown of rays and a team of fiery steeds, gradually receded in importance until most of his functions were absorbed by the now more cerebral Apollo, god of purification, healing, and rationality—no longer the plague-shooting marksman of the *Iliad* but instead the soteriological deity of the Pythagoreans and neo-Pythagoreans. The culmination of the Greeks' theomorphism was the glorified human forms of the Olympians as they appear on the inner frieze of the Parthenon and in monumental chryselephantine temple statues. Later in Jewish and Islamic traditions, as the writings of Maimonides and Al-Ghazāli show, anthropomorphism had to be rejected in orienting oneself to God as definitively as Abraham rejected the polytheistic star worship of Terah, or the Prophet rejected the worship of the moon daughters of Allāh. Rudra, "the Roarer," the Vedic storm god, receded in importance, as later Brāhmaṇical and Upaniṣadic texts emphasized and then philosophized the divine status and powerful autonomy of sacrifice itself, its sacred metrics, instruments, and victims. Other, more "human" gods, with complex cosmic and moral ranges, survived as figures of influence in this later period of Vedic thought. Gods whose main purview was nature, like Rudra and Sūrya, became vestigial.

The work of scholars on the central role of natural symbolism in religious systems has now been "problematized" to the point of irrelevance by religious studies' sociological and political emphasis. But the definition of ritual offered by the Africanist Evan Zuesse, remains to challenge such chronologies in which religions progress from "less developed," "nature-oriented" traditions to highly developed, "abstract" ones completely divorced from the natural environment. At the "developed" end of the scale, entities like lightning or animals exist only in highly stylized, symbolic forms as functions of divine creativity or sacred history (as in the menorah's obvious heritage in the tree, or perhaps even in the *asherah*).[5] Zuesse defines ritual as "those conscious and voluntary, repetitious and stylized symbolic bodily actions that are centered on cosmic structures and/or sacred presences."[6] The fact that in any tradition, cosmic structures and sacred presences are just as likely as not to be located in the natural, observable world requires that the study of religious thought and praxis rectify its tendency to regard such locations as primeval or underdeveloped.

On the other hand, many traditions do *not* follow this developmental train of progression. Ancient Egypt is a noteworthy example. Not only did theriomorphic gods exist as central players in an esoteric and self-referential religion of magic, astronomy, personal accountability, and excruciating eschatology, but also through its distinctly solar theologizing, Egypt produced the first known instance of monotheism in world history during the short spasm of Akhenaten's reign, when the sun became a perfect vehicle for the concentration of divine power and the supernatural dispensation of justice and fertility. In one of the world's most complex polytheistic systems, the Hindu Gaṅgā is both goddess and river. It is not that the goddess dwells in the river or that the river "symbolizes" something about the goddess. Rather, she *is* the river, so her body is sacred and supremely pure and purifying. Gaṅgā is the greatest *tīrtha*, a crossing place or ford into the next world, a bridge between immanent and transcendent realms.[7] Among indigenous traditions, most of which relied on oral transmission and were not literate until relatively recently, the so-called worship of nature is almost never that, but instead describes the natural world in an intricate, heavily determined web of relationships, a web that in turn maps multivalent forms of religious and cultural thought. Consider, for example, the Australian aboriginal idea of the Dreaming, in which ancestral dream heroes—Kangaroo, Rainbow Serpent, Barking Spider—created features of the natural landscape long ago, yet through human cultic maintenance also indwell or even incarnate that same landscape—or the Maori dichotomy between Tane, the personified earth, and Tangaroa, the personified sea, caught up between themselves in an electric volley of power, betrayal, and mutual succor, along with unexpected, intense roles in human intrigues.[8] Or think on the Inuit theologies of *tuunraq*, helping spirits, and *inuit*, the "free souls" in all beings, including animals and entities like the sea, who express complex moral range as well as divine polarities. Western concepts like "symbolization" do not begin adequately to illumine such systems, yet their main players, their sacred presences, are semiotically bound up with the physical environment, the sensible natural world and its forces.

Natural Purifying Features of the Sea, Religiously Constructed

In chapter 2, we considered the features of the marine environment and oceanic properties that make the sea a natural global filtration system. In this chapter, we distinguish some of those major marine features that have been "taken up" in the religious imagination, focusing on those that have seemed to be crucial to the sea's role as a vehicle of catharsis.

Water

Purification by water has been and remains at the center of numerous religious rituals throughout the world, particularly those signifying metamorphosis: revolutions in the life cycle, movements from one condition to another.[9] Part of the reason for this is that water seems to be a primary signifier for metamorphosis and metamorphic potential, itself somehow not subject to change because it is the essence of change—at least in the religious imagination. Theodor Schwenk observes that water "does not grow because it is itself the universal element of change that runs through every possible form without becoming fixed in any."[10]

For example, although the sea's ritual purifying powers surely are symbolically related to those of any smaller freshwater *miqveh*—indeed, the sea, if available, surpasses the *miqveh* for daily purifications, for the *miqveh* symbolically represents the sea, not the other way around—marine ritual properties are far more chronic than those of the microcosm.[11] In those cases, divine destruction and recreation also are effected through water. The sea was the first *miqveh*, the "gathering" of waters below, separated from the waters above as God created the world in Genesis: "And God called the dry land "earth," and to the *miqveh* of water he gave the name *yamim* [sea]" (Gn 1:10).

The waves, tides, and currents of the sea seem to multiply ordinary water's power infinitely. The sea is not a static, artificial container of purification that must be created (like a *miqveh* or a baptismal font, where waters are transported and consecrated) but is a watery entity, alive and available. In its omnipresence, it makes chronic cleansing possible. The sea, itself ultrapure, can receive and clean what is also too pure or holy, too "charged" to be flushed into the city's drains and sewers. Since baptismal garments are worn only once in Eastern Orthodox Christian tradition, made or purchased new with each baptism to symbolize the death and recreation of the unique soul in the water of the font, new generations of infants do not wear heirlooms. Nevertheless, the baptismal garments and white towels in which the baby was wrapped when she emerged from the water are preserved. On Greek islands and along coastal areas, such things are never cleaned at home but are taken after the rite to the sea. There in the waves, and only there, one may wash from them the consecrated oil that was added to the water in the font during the sacrament, so that the well-greased baby might "slip through the devil's grasp." Inland, the garments are often burned, fire and seawater sometimes being corollary means of ultimate purification, as we will consider later.

The waters of the sea scour; they erase any trace of the signifiers that render artifacts human or sins eternal. "You will hurl into the deepest sea all their trespasses," says the prophet Micah to God (Mi 7:18–19). The idea is that in the depths of the sea, wrongdoings will be lost forever to human, and even to divine, memory. In the annual ritual of *tashlikh*, held since medieval times on the first afternoon of Rosh Hashanah, or on the second day if the first falls on Shabbat, personal sins are carried to the sea in the form of bread crumbs and washed away forever.[12] As with every sacrificial offering, both before and after the destruction of the Temple, intentionality is stressed, and is in fact what determines the efficacy of the rite. The dropping of the bread crumbs into the sea is not enough to expiate a year of sins; instead, repentance must inform the action (this is the main theme of the Hasidic wonder tale "Gershon's Monster," discussed in chapter 7). Lesser bodies of water may substitute for the sea in *tashlikh:* a river or stream if necessary, although, as in the ancient Greek typologies of purification, running water is preferred. But like the *miqveh*, the ritual bath, these "stand" in some way for the paradigmatic cleansing ocean. For evil to enter the sea, then, is for it to be existentially changed, lost into oblivion: the sins of the previous year, the pursuing chariots of Pharaoh in the Exodus story, the demons driven by Jesus into the swine. All these the sea swallows forever.

In the Talmudic tractate *Baba Mezia*, at 22B and 24b, we read that if one finds something that has fallen into the deep sea, one can keep it. Moses Maimonides later comments in his Mishneh Torah, "If you find a lost object at the bottom of the sea, even though it has markings on it clearly identifying to whom it belongs, the finder can still keep it." It is assumed, in other words, that the owner relinquishes his or her property rights when the object plummets down.[13] The condition of being in the deep sea thus erases the former marks of attachment or possession, which otherwise would compel the finder to search for the original owner. What falls into the sea is lost forever, and what comes out is forever transformed. It is not surprising, then, that in the ancient religious imagination, this physical property of the sea to clean and to change became so metaphysically charged.

Salt

The symbolic efficacy of salt, the second most prominent of the sea's constituent elements, is affirmed in countless traditions. Salt preserves cell-based organisms, both food and the bodies of sacrificial victims, from which it can absorb blood and render the flesh pure, as in the cultus of the Jerusalem Temple. Saltwater also corrodes as nothing else can, even breaking down metals.

The salinity of the ocean varies according to cycles of ice buildup and melting in glacial periods.[14] Two hypotheses account for the ocean's salinity: the first is that the ocean was a freshwater lake that became gradually salty. This theory accounts for the saltiness of the ocean's water by arguing that for eons rivers have been adding dissolved salts to it. By analogy, the Dead Sea and Great Salt Lake have a high saline content because they absorb incoming dissolved salts at the same time as their water is simultaneously evaporating. The second hypothesis holds that the sea was always salty to begin with: when the molten earth cooled, silicates, which have a higher melting point than chlorides, solidified first and left their salt on the surface. The earth's further cooling resulted in a condensation of water vapor, and the salt dissolved in the global waters that are now the seas.

Salt has long been associated with purity and purification of every kind. In Diogenes Laertius's life of Pythagoras, the sixth-century B.C.E. philosopher, mathematician, and mystic, one Alexander, a maven of Apollonian purity, cites his teacher on salt: "It should be brought to table to remind us of what is right; for salt preserves whatever it finds, and it arises from the purest sources, sun and sea

(ἐκ τόυ κατηαρōτάτōν ηέλίōυ καί τηαλάσσέ)."[15] The eighth-century
C.E. Japanese text *Kojiki* preserves the myth that while creating the
universe, the god Izanagi made the first pure island of Onogoro out
of salt that he took from the primordial sea. By extension of its pres-
ervational and purifying character, salt is often also ascribed apotropaic powers, not only to neutralize malignant forces but actively
to repel them. Elisha threw salt into bitter waters as an exorcism (2
Kgs 2:20–22); speaking eschatologically, Jesus told his disciples, "For
everyone will be salted with fire" (Mk 9:49).[16] The sixth-century C.E.
Roman Christian *Gelasian sacramentum* tells how salt drives away the
devil from catechumens.

Infinity

The apparent inconsequentiality of anything that is put into the sea
is due to the latter's corresponding apparent boundlessness, reflecting, like a troubled mirror, the infinite vault of the sky. The sea's apparent extension to the very edge of the horizon was, of course, what
led to the illusion of the earth's flatness—challenged by the ancient
Greeks as they observed the circular shadow cast by the earth on
the moon during a lunar eclipse and rechallenged millennia later by
the voyage of Columbus. And that is only the view from the land.
When one is "out to sea," one has an overwhelming sense of its oc-
cupying every quadrant of the four directions and more, filling the
horizon in a circle and creating the ultimate theater for pilgrimage,
against which the human boat is, like something thrown into the
waves, infinitesimal. The Celtic poem attributed to King Cormac
mac Cuileannáin (d. 903) mediates upon this contrast, when the
seafarer asks Christ,

> Shall I take my dark little coracle
> on the broad-bosomed glorious ocean?
> King of the shining kingdom, shall I go
> of my own will upon the sea?[17]

The sea is the vast and fearsome "circle" surrounding novelist Yann Martel's young Indian castaway Pi and his accidental companion, a Bengal tiger named Richard Parker, in *Life of Pi*. Pi's story is described by one of its characters as "a story that will make you believe in God," as the heroes drift 7,000 miles in an open lifeboat.

> To be a castaway is to be a point perpetually at the centre of a circle. However much things may appear to change—the sea may shift from whisper to rage, the sky may go from fresh blue to blinding white to darkest blue—the geometry never changes. Your gaze is always a radius. The circumference is ever great.[18]

Perhaps because of its contrast with tiny human frailty, marine infinity often serves as a metaphor for the vastness of what is precious, universal, or divine. The teacher or the teaching becomes the ocean, or the ocean becomes the teaching. The Vietnamese monk Thich Nhat Han remembers the first sermon, "The Buddha said that the water in the four oceans has only one taste, the taste of salt, just as his teaching has only one taste, the taste of liberation."[19] For centuries rabbis have referred to "the ocean of the Talmud." The great Sufi mystical poet Jalāluddīn Rūmī addressed the Prophet, "O you Mercy for the worlds! Out of the ocean of certitude / Do you grant pearls to the dustborn, peace to the fishes!"[20] Shamsuddīn Aḥmad Aflākī wrote in the fourteenth century of how Rūmī interpreted the symbolism of a man's dream of the sea, saying, "That endless ocean is the Greatness of God most High."[21]

The infinite, all-encompassing persona of the sea has also made it a ready simile for the Freudian psychoanalytic equation between the putative human experience of parental love in early infancy and the (equally putative) derivative experience of adult religiosity. The infant's experience is one of "primary narcissism," before the encounter with a recalcitrant, unyielding reality that does not automatically affirm her every wish:

> The infant's early experience is one without boundaries: there is no boundary between the child and the world; instead, the

child experiences the mother's love for it directly and without mediation. . . . The infant thus experiences itself as being continuous with the world, with a loving world . . . yet if we reflect on it, we may discover that in it lies the kernel of some religious experiences in later life, when the religious believer loses him or herself in a religious community, experiencing an "oceanic" feeling of love, that once life was perfect, as in the Garden of Eden . . . we are tapping into this substratum of our experience.[22]

Here the ocean becomes a metaphorical bridge between the "primordial" and paradisiacal state of infancy and its purported recapitulation in the erasure of ego boundaries and unconditional love in religious experience.

Depth

"All the rivers run into the sea, yet the sea is not full" (Eccl 1:7). The illusion of the sea's endless depth, and hence its ability to contain all that it receives, is made not only by the submerged landscape of its black chasms, ravines, and valleys but also by the fact that efforts to find its deepest point have repeatedly failed.

At its nadir, the sea is not only cold and utterly lightless; it is also subject to pressure so great as to be lethal to human beings. Some species use bioluminescence to locate one another in the dark or to repel predators; many others use sound, transferring acoustic energies in ways we cannot accurately measure. Just below 3,000 feet off Georges Bank, where no light penetrates, soft-water and hard-water corals intertwine in colonies of vibrant, apparently pointless color. They have names like "bubble gum coral," "strawberry coral," "gold band coral." Since corals grow less than an inch a year, one sweep of a deep-sea trawler sets back by decades their growth and the viability of the myriad organisms they are thought to harbor.[23] And even at 8,000 feet or deeper, as marine scientists first discovered off the Galapagos archipelago in 1977, there is life, of the chemosynthetic rather than the carbon-based kind, where giant clams and tube worms, along with blind, colorless crabs, dwell by hydrothermal vents spewing sulfur. The recently discovered microbial *Archaea*, "ancient ones," who also inhabit these vents and chimneys, are now believed to have been the first forms of life on earth.[24]

Moby Dick, one of literature's most mysterious main characters—and a nonhuman one at that—in one of America's most important religious novels—was a sperm whale, a denizen of the deep. Sperm whales are rarely seen near shore on whale watches, as they roam in deeper waters, consuming a ton of fish and squid a day with teeth five inches or more in length, which easily can snap a human leg or crush a wooden boat. They dive to depths of more than half a mile, into the great oceanic chasms such as those off the Mid-Atlantic Ridge. Sperm whales can stay submerged for more than an hour, "and they can swim at a speed of twelve knots, easily outpacing rowed whaleboats. . . . They are, in effect animals of two worlds: they know the murky unlit mysteries of the deep; but they're also air-breathing mammals that know the look of the open sea and sky."[25] Moby Dick, a figure of both divinity and evil in the novel, is a creature whose symbolic profile corresponds to the extremity of his home—beyond human power to reach or even to apprehend.

As Catherine Keller observes, though, targeting the crux of the matter: "Does the bottomless world-source [the sea] suggest bottomless *resources?* To the contrary; what materializes from its matrix, what actualizes, is always and only *limited.*"[26]

Antiquity, Eternality, and Chaos

To both the formal religious mind and the speculative casual observer, the ocean appears to be ancient and eternal. It seems to have "been there," restlessly rolling since the beginning of time or perhaps before. The Atlantic Ocean is 200 million years old, but the global sea is much older, the vast relic of unimaginable rains as the earth's surface cooled 4 billion years ago; so an environmental scientist like

Rachel Carson, a denizen of the American East Coast, could write, "To stand at the edge of the sea, to sense the ebb and flow of the tides, to feel the breath of a mist moving over a great salt marsh ... is to have knowledge of things that are as eternal as any earthly life can be." It is perhaps no wonder that biblical theologians as different from one another as Jon Levenson and Catherine Keller have pointed to the traces in Genesis and elsewhere of the notion of the sea's preexistence and autonomy from established divine systems. In this ancient Near Eastern idea, the sea represents the prehistoric embodiment of formlessness and chaos that continues to threaten order even after it is divinely established, as one can see in the parallel ordeals of Marduk and Timat, Baal and Yamm, or Yahweh and Leviathan.[27] In nearly every narrative in which the ancient sea "pre-exists" the rest of the natural world and all later human cultural forms, it tends to symbolize the opposite of culture itself.[28] As is often the case with other primordial forces in myth, Ocean is construed as wild, unruly, and undifferentiated *because* it is timeless. Its chaotic nature is somehow implied by the facts of its endless past and equally endless future. Apparently nothing so old can be imagined as ordered.

Mutability and Magical Quest

The sea's continually shifting planes and its mercurial responsiveness to weather patterns have made it a religious emblem of mutability. The waves of the sea are the arena of hallucinogenic shape shifting, of wild transformation. They have neither fixed planes nor fixed levels and thus seem to spawn an infinite number of conditions. As Yann Martel's drifting Pi says,

> There were many seas. The sea roared like a tiger. The sea whispered in your ear like a friend telling you secrets. The sea clinked like small change in a pocket. The sea thundered like avalanches. The sea hissed like sandpaper working on wood. The sea sounded like someone vomiting. The sea was dead silent.[29]

By association with their matrix, oceanic creatures are often slippery, wise, prophetic, or magical, like Long Wang, the Chinese dragon king with a human body who lives under the sea; like Mahākāla, serpent-king of the submarine *nāgas*, those ancient magical ones who zealously guard in splendor one of Doṇa's original portions of the Buddha's relics after they were washed away in a great flood from the Rāmagāma shrine;[30] or the wily Old Man of the Sea, Proteus, with whom the Spartan king Menelaos wrestles in the *Odyssey* seeking knowledge of the lost king Odysseus as the Old Man turns into a lion, a serpent, a boar, fluid water, and a tree.[31]

The ocean that encircles the world is the arena of quest and personal transformation. While the Greek *Odyssey* gave us a generic name for the initiatory journey, it cannot account for the many other sea odysseys—the magical Celtic *imram* tales, for example, including the eighth-century C.E. *Voyage of Bran*[32] or its tenth-century Christian successor, *The Voyage of Brendan,* with its monastic islands made of crystal and a great monster named Jasconius who slowly glides in concentric circles as unwitting seafarers build bonfires on his back. Here, then, the mutability of the sea resonates with what it often offers its voyagers: radical change.

Catharsis and Containment

Because of all these properties, the sea offers ultimate catharsis and the related idea of renewal or even resurrection. This aspect of the sea is surely related to the group of flood myths in which divine destruction and recreation also are effected through water already upon the earth, pouring from the sky, or both: the deluge sent by Heart of Sky to destroy the divinely annoying wooden manikins in the Mayan *Popul Vuh,* or the flood that cascades from heaven in Genesis 6 to relieve God's heartache at the evil that he witnesses,[33] or the cataclysm transcribed by Sin-Leqqi-un-ninni in the eleventh tablet of the neo-Babylonian text of the *Gilgamesh* epic, a divine deluge so horrible that "the gods cowered, like dogs crouched by an outside wall."

The cathartic sea effects change of a special order, and nothing comes out of the sea the way it went in—not trees, not boats, not corpses, not ritually impure or guilty people—not even, as we shall see, the fire that signals the end of the triple worlds.

Burial of the Dead

The sea is a birthplace; it is also a graveyard. Or it may be the passage from the Grey Havens to the far green shore, the place of the "swift sunrise." The idea of the sea as the arena of a great death-journey lies behind late Old and New Kingdom Egyptian ship biers, as well as the prehistoric boat-shaped coffins discovered in the burial mounds of pre-Buddhist Japan. The *Sui shu,* a seventh-century Sui dynastic history, offers textual corroboration for the ancient Japanese customs: "For burial the corpse is placed on a boat and sometimes rollers are used to pull it along the land."[34]

On Memorial Day at Ala Moana Beach in Honolulu, in the Japanese Buddhist ceremony of Lantern Floating (*Toro Nagashi*), the bereaved launch more than a thousand lit paper lanterns onto the night sea. The small glowing shrines float on carved rafts, some with masts and gossamer sails; they bear the spirits of the dead over the horizon "from the shore of delusion to the shore of salvation."[35] Like the ancient Greek Isles of Blessed, like Lewis's Land of Aslan or Tolkien's far green shore, our final home—the place where our spirits will sail, once freed from our bodies—lies on the other side of the ocean.

During the late Iron Age in Scandinavia, from around 400 to 1050 C.E., at the Gokstad and Oseburg mounds, royal funerary ships show how oceanic burial rites were combined with eschatological beliefs. As the slain god Baldr receives a fiery funeral at sea in Snorri Sturluson's *Gylfaginning*,[36] so the Old Norse chieftain's serpent-headed boat was his sarcophagus in the earth, along with his horses, dogs, bedsteads, and sledges—and, in the case of Viking burials in the Hebrides and along the Volga in Russia, his female servants.[37] The Salish Indians of the Pacific Northwest coast buried their dead in canoes suspended by poles, paralleling Polynesian practices; a Twana tale tells how denizens of the realm of the dead could be heard paddling along the coast at night to claim the newly deceased.[38] The funeral and burial at sea became commonplace on European sailing ships; for the sea itself was a sort of great coffin. That this idea can be revived very quickly in the event of a marine disaster can be seen from the history of public response to the sinking of the *Titanic*, up to the time of its controversial underwater "recovery," or the changed responses to the now charne sea by the Indian Ocean survivors.

The Sea as the "Ground of Being"

Eliade's "structure and morphology of the sacred," expressed in his systematic vision through the potent and universal hierophany of water, finds perhaps its strongest form in the ocean. "Every sacred space implies a hierophany, an irruption of the sacred that results in detaching a territory from the surrounding cosmic milieu and making it qualitatively different . . . something that does not belong to this world has manifested itself."[39] As Smith puts it, "This manifestation ontologically grounds (or "founds") human existence." As the uninhabitable, encircling matrix for the inhabited world, the religiously imagined sea—a forceful expression of "space"—is a constant. It is the reference point for our land-based existence, nourishing that life at the same time as it threatens it and carries off its poisons. The sea, groundless, is our ubiquitous "ground of being."

But the Eliadian category of "hierophany," insofar as it implies the expression of something uncanny in something ordinary, is, in a sense, undermined by the nature of the sea itself. In myth, the sea, older than memory or culture, older than time, older than the cosmic ordering actions of the divine, existed before "the world." The sea thus does not reveal a foreign, hierophanic force that "does not belong" to this world. It is not a territory "detached" from its cosmic milieu by selective manifestation but rather is a territory already aquatically self-defined and believed to be inherently efficacious.

In most mythic cycles and religious traditions, the sea does not "index" something greater than itself, except in the monotheistic protest that it does indeed obey God. But why is the protest repeated so forcefully in sacred histories? In its antiquity and continual newness, in its familiarity and utter unfamiliarity, in its generativity and destructive potential, the sea seems to contain all the paradoxical binaries of the transcendent. It "stands for" and manifests only itself. Hence, it apparently has the power to dilute and thus transcend what has gone amiss on land. The next three chapters show how this occurs in similar and yet radically different traditional religious settings.

CHAPTER 4

"The Sea Can Wash Away All Evils"

Ancient Greece and the Cathartic Sea

THOAS: *How shall we deal with the strangers?*
IPHIGENIA: *We must respect the law of the sanctuary.*
THOAS: *Your pitcher and great knife, then, for the sacrifice?*
IPHIGENIA: *First I want to immerse them, for the sake of holy purifications.*
THOAS: *In the waters of springs, or in the salty sea?*
IPHIGENIA: *The sea can wash away all evils of humankind.*
THOAS: *It might be more sacred, then, to the goddess.*
IPHIGENIA: *I myself think it is better.*

Euripides, Iphigenia in Tauris 1188–1195[1]

Because of its geophysical situation, Greece is home to a people for whom the sea has been a potent religious, aesthetic, and economic entity throughout history. The Mediterranean is an eternal presence in the landscape, charged with cultural meaning for a society that was oriented as much to the sea as to the land from its earliest stages. This reality was enacted in the opening ceremonies to the Olympics in Athens in the summer of 2004, when the enormous train of a tiny Icelandic singer's aqua and green dress, representing the ocean—central to Greek imagination and identity—gradually unfolded, billowing, to cover the entire theater of the stadium and all the athletes from around the world beneath it. Seagoing embassies and battles are depicted in the wall paintings from Thíra from around 1500 B.C.E., during the period of the Minoan thalassocracy; marine fauna somersault across the sides of Mycenaean chamber tombs and Minoan sarcophagi.[2] The meaning of this marine iconography in early funerary contexts is unclear, but the choice of theme cannot be accidental and must somehow be related to personal or group eschatology.

From at least as early as the archaic age, the dangers of navigation ensured a series of traditional ritual acts associated with seafaring: libations and propitiatory sacrifices for controlling the weather and calming storms.[3] No large boat in antiquity sailed without an altar on board, and the special realm of nautical safety even had its own mystery cult, that of the Samothracian Kabeiroi and Megaloi Theoi.[4] The Aegean Sea was nominally ruled by Poseidon, brother of Zeus (originally the god of earthquakes but later acquiring the realm of the deep as well) and his Nereid sea wife Amphitrite, whom Theseus visited on the ocean floor on his way to Crete to battle the curse of the Minotaur. But the sea had countless other faces, and countless forces to appease. Like briny alphabet soup, it was crammed with a legion of divine and temperamental figures, such as the shape shifters Phorkys and Proteus; Nereus and his fifty daughters; mermaid figures like Thetis, the mother of Achilles; or Ino Leukothea and her mysterious veil of salvation that floated the drowning Odysseus to Phiakia.

Death, Rebirth, the Sea, and the Ocean

Mainstream Greek notions of purity underwent an extraordinary metamorphosis from the Geometric period in 800 B.C.E. to the late classical period in the 400s. Earlier, "purity" was what one might call a material condition. Even at Delphi, no real notion of moral repentance or interiority was attached to the cleansing of blood or wrongdoing until the late classical period. In other words, the archaic Greek human being was purified by what he or she did about the stain of pollution, not by how he or she felt about what had initially caused it.

There was an outstanding strand of interwoven exceptions to this chronology, and it appeared very early, in the sixth century B.C.E. An emphasis on physical purity as a means to psychic rebirth, itself in turn a means to ecstatic immortality in the afterlife, shared with

the gods, is well in known in Pythagorean, Orphic, and Bacchic esoteric teachings. The goal was an escape from the normal gloom and
meaninglessness of Hades. Nevertheless, it is possible to argue that
the overlay of purification and its associations, the cleansing of sin
and death, and deathlessness in the mysteries of the archaic period
had precedents, no matter how fugitive, in earlier Greek symbolism
of the sea—the ultimate source and producer of ritual purity and
thus its moral correlative.

In our main source of Greek cosmogonic myth, Hesiod's *Theogony*, the salt sea, Pontus, is sharply distinguished from Okeanos,
a great river of fresh, sweet water encircling the earth's flat disk,
floating upon primordial waters, and forms the outer limits of the
world. Both Pontus and Okeanos are the children of Gaia, who
came into existence just after Chaos. But Pontus, the open sea, the
high sea, is the parthenogenetic child of Earth, born "without de-
lightful love," and is himself appropriately fruitless, *atrúgeton*: "the
sterile main, raging in its swell" (*oídmati thuîon*, *Theogony* 131; *thuîon*
comes from a verb used in the wild raving of Dionysiac maenads).
"Deep-eddying" Okeanos, though, is the child of Earth and her
own offspring Heaven (Ouranos), who is a suitable mate because he
is "equal to herself."

As discussed in chapter 3, the theological situation in Greek religious
history is complex at this point, around 700 B.C.E. These gods of nature
are *gods* and should not be understood otherwise. As Jean-Pierre Vernant
writes in his discussion of Greek cosmogonic myth, Hesiod's account
"opens with the evocation of the divine powers whose names, places,
and roles indicate their cosmogonic signification. These 'primordial'
gods are still enough a part of the physical realities which they evoke
that we cannot separate them from what today we could call forces
or elements of 'nature.'"[5] That they are not anthropomorphic does
not primitivize them or mean that the Greeks thought that their own
cosmogonies were "really" about elements of nature as such, only
"thinly disguised" by divinization. Instead, it means that an extra
measure of effort is necessary to comprehend not only the genealo-

gies but also the essence and motivations of such "nonhuman" deities of antiquity.

The *Odyssey* has Pontus as its theater of action and transformation. Homeric epithets speak of the salt sea's enormity: "wide back," "wide lap," and "wide paths"; it can be "gray-blue" or "wine-dark"; sometimes, as we have seen, it is *atrúgeton*, "barren," but at other times *ichthuóenta*, "full of fish."[6] But the *Iliad* is far more interested in the great freshwater river Okeanos, which in this epic reveals its cosmic, generative, and regenerative nature. In a sense, the dimensions that often coalesce in religious conceptions of the sea are bifurcated in the Greek religious imagination into these two great bodies of water. Pontus bounds the land, and in its wildness is a world and a law unto itself. But freshwater Okeanos bounds all of known cartography, since it is encountered only by traveling to the edge of the salt sea, just as C. S. Lewis's *Dawn Treader* must sail to ocean's end to find the sweet fresh waters and lilies of Aslan's country. When Zeus summons a divine assembly on Mount Olympus, even the river gods and spring nymphs come: "Okeanos alone remains at his station,"[7] for Okeanos both defines and sustains cosmic limits. It is where Helios himself plunges every night, as when "the shining light of the sun was dipped in Ocean" and reemerges at dawn, when it rises "out of the quiet water and deep stream of Ocean / to climb the sky" (*Iliad* 14.485 and 7.421–422).[8]

Because of its role and location in the quotidian solar drama, as Gregory Nagy discerns, Okeanos is also the quintessential place of death and rebirth. Odysseus begs Artemis to end his life, carrying him "down misty pathways, / and set me down where the recurrent Ocean empties / his stream; as once the stormwinds carried away the daughters / of Pandareos" (*Odyssey* 20.64–67). Yet Okeanos is also, astonishingly, twice called *theôn génesin*, "source of the gods," in *Iliad* 14.201 and 302, and even *génesis pántessi*, "source for all things" in *Iliad* 14.246. Okeanos is thus aligned with what Nagy calls "re-animation" (from the verb in *Odyssey* 4.568, *anapsúkhein*, literally, "to ensoul again") and immortality. Even though the shore of Okeanos is the place of the entrance to Hades, where Odysseus must sail and descend as no living man has done before him to receive his homecoming prophecy from the ghost of Teiresias (*Odyssey* 11.20–21), it also is the place of the Elysian Fields (*Odyssey* 4.567–568), where chosen heroes spend the afterlife in warmth and beauty. There Okeanos sends up "gusts of shrill-blowing Zephyros / at all times, so as to re-animate human beings" (*anapsúkhein anthrópous*).⁹ Hesiod's Works and Days (171–172) tells us that along its shores lie the "Islands of the Blessed," the home of the godlike race of heroes who ruled the earth before the present generation, "untouched by sorrow"—or death—and ruled over by the exiled Kronos.

But it is not only the mysterious Okeanos who flows with the streams of death, rebirth, and immortality. Raving Pontus, perhaps by watery extension, shares too in these thematic currents; or we might ask whether eternal and rarely seen Okeanos is not an idealized expression of Pontus's better qualities, set by myth at the ends of the world. The "real" ocean, across which lay scattered the Mediterranean world, island and intricate coastline, was obviously a place of peril and frequent loss. Yet through its restless sarcophagy, it preserved civil life's status quo even as it dealt death to the unfortunate, consuming their bodies in its maw. Chapter 3 introduced the idea that the salt sea's cathartic powers are underscored by—or perhaps ultimately originate in—its ability to swallow and dissolve what is often the most extreme object of pollution, the most problematic object in any society: the corpse. In the Greek Bronze Age, continuing at least through the Geometric period with its visceral, angular confrontations in art with themes of death, mourning, and the mutual slaughter of men and monsters, this power often was expressed with the image of hungry fish. In the epic poems, the sea is *ichthuóenta*, "full of fish"; thus the myriad denizens of the sea, the fish, allow it to "eat" corpses, which it then refuses to surrender.

The sea consumes and therefore negates the charged pollution of death but, in doing so, aborts perhaps the most important cultural container of meaning in ancient Greece, the signifiers of a life well lived: a recovered corpse and a decent funeral. It was the ironclad, sacred obligation of one's family and friends, those at the top of what Gregory Nagy termed "the ascending scale of affection," to cremate their dead in a splendid pyre (in Homeric idiom, but actually reflecting Geometric funerary practices) or bury them piously in the soil of one's home (in the archaic and classical periods and later)—and to offer heartfelt lamentation. The sea could take all that away as well, and it did. Thus Achilles curses his battle victim Lykaon, son of Priam, with a fate much worse than death:

> Lie there now among the fish, who will lick the blood away
> from your wound, and care nothing for you, nor will your mother
> lay you on the death-bed and mourn over you, but wide
> Skamandros
> will carry you spinning down to the wide bend of the salt water.
> And a fish will break a ripple shuddering dark on the water
> as he rises to feed on the shining fat of Lykaon.
>
> *Iliad* 21.122–127[10]

As Emily Vermeule observes, Iliadic fish are "raw-ravening," like birds, dogs, lions, and Achilles himself.[11] Indeed, a Geometric-period vase, dated at the same time as the collation of the Homeric epics, from Pithecousai in the Ischia Museum, depicts fish ready to devour each one of the hapless victims of a shipwreck; one large fish has already caught someone's head in its jaws.[12] Pontus was no romantic entity for the ancient Greeks but a comfortless place of ordeal and wildness, where ritual cannot be properly observed, nor the gods well fed. Nevertheless, it retained its genealogy and its divine autonomy.[13] The nonmarine gods feel this as keenly as mortals, as when Hermes flies down to release the weeping Odysseus from the clutches of the nymph Kalypso on her enchanted island, deep in the heart of the sea. The winged-sandaled one complains as bitterly as any unwilling business traveler:

You, a goddess, ask me, a god, why I came, and therefore
I will tell you the whole truth of it. It is you who ask me.
It was Zeus who told me to come here. I did not wish to.
Who would willingly make the run across this endless
salt water? And there is no city of men nearby, nor people
who offer choice hecatombs to the gods, and perform sacrifice.

Odyssey 5.97–102

The salt sea is not only a bleak arena of death and danger. Strange
ideas about rebirth and immortality inhered there, too, again beginning as early as the epic imagination, although they are revealed far
more graphically in the realm called Pontus than they are in the
symbolic field of Okeanos. By the time he falls into the clutches of
the sea, after his release from Kalypso, Odysseus has already made a
grim, unorthodox descent into hell. And he has reemerged, like the
sun.[14] He has acquired knowledge denied to most mortals, namely,
the time and manner of his own death, as the blind shade of Teiresias predicts that "Death will come to you from the sea, in / some altogether unwarlike way, and it will end you / in the ebbing time of a
sleek old age."[15] Thrust back then into the sea god's realm, though,
Odysseus is nearly destroyed "beyond his destiny" (for such a thing
was possible in the ancient Greek conception of Fate; something
could occur, rarely, yet entirely possibly, *hupèr móron*, "against
fate," something that not even Zeus could override). His raft is torn
apart by the wrath of Poseidon as the god "staggers the sea" and
sends up enormous waves against him. He who throughout his odyssey has been *polútropos*, "turning in many ways; ingenious"—like
an octopus, with its constant defensive color changes—cannot help
himself now.

Unearthly aid appears to the hero in the form of the mermaid goddess Ino Leukothea, "the white goddess," who, like the Inuit Sedna,
was once a mortal woman who suffered a tragic plunge into the sea.
A Theban royal, daughter of Kadmos, Ino was driven mad by Hera
as punishment for nursing her own nephew, the infant Dionysos,

after Zeus's lightning killed his unfortunate human mother, her sister Semele.[16] But Ino did not die; rather, "now in the gulfs of the sea she holds degree as a goddess" (*Odyssey* 5.335), who was worshiped throughout the ancient Mediterranean and perhaps had connections to the Syrian fish goddess Atargatis.[17] The seaborne immortal Ino Leukothea has a magical veil with lifesaving properties, and taking the form of a sea bird, she gives it to Odysseus to keep him from drowning, urging him to leave his clothes and the raft behind and swim for shore. "And here, take this veil, it is immortal, and fasten it under / your chest; and there is no need for you to die, nor to suffer" (*Odyssey* 5.346–347). After skeptically wondering whether her plan could possibly be a trap, Odysseus has things decided for him when a massive wave sent by Poseidon crashes down on him. He strips, the veil of Ino around his chest, and strikes out for shore, but encounters deadly surf against the cliffs of Phaiakia:

> . . . a great wave carried him against the rough rock face,
> and there his skin would have been taken off, his bones crushed together,
> had not the gray-eyed goddess Athene sent him an inkling,
> and he frantically caught hold with both hands on the rock face
> and clung to it, groaning, until the great wave went over. This one
> he so escaped, but the backwash of the same wave caught him
> where he clung and threw him far out in the open water.
> As when an octopus is dragged away from its shelter
> the thickly-clustered pebbles stick in the cups of the tentacles,
> so in contact with the rock the skin from his bold hands
> was torn away. Now the great sea covered him over,
> and Odysseus would have perished, wretched, beyond his destiny,
> had not the grey-eyed goddess Athene given him forethought.
>
> *Odyssey* 5.425–437

Odysseus swims out to open sea and around the island until he spies the mouth of a river, to whom he prays:

Hear me, my lord, whoever you are. I come in great need to you,
    a fugitive from the sea and the curse of Poseidon;
even for immortal gods that man has a claim on their mercy
who come to them as a wandering man, in that way that I know
come to your current and to your knees after much suffering.
Pity me then, my lord. I call myself your suppliant.

                                        Odyssey 5.445–450

The river hears Odysseus' prayer and stops the deadly waves. Odysseus crawls out of the brine onto the shore of Phaiakia, naked, spent, his ship and all his men long gone. This will be his last stop before home and the place where he will be sheltered and helped. Phaiakia is where he finally grieves the Trojan War, hearing the bards' unwitting feast songs of all he has personally endured. It is where he at last is provoked to recount publicly the story of his wanderings. The ocean's terrific assault, an instrument of Poseidon's wrath, has stripped Odysseus not only of all his possessions but also of all his markers of identity. Having plunged into and emerged from Okeanos, the world-circling stream, Odysseus has undergone a cosmic death and resurrection. He has made a trip into Hades, cross-examined the blood-drinking shades, among whom are his mother and the archetypal *hérōs* Achilles, and learned the manner not only of his homecoming but also of his own death. His encounter with Pontus repeats these themes, but in a literal way, as the sea batters him with its great waves, nearly drowning him, and shattering all that made him who he was—violently pulling the octopus from its rocky shelter. The encounter with Okeanos, at the edge of the known and navigable world, is one full of eschatological peril; the struggle with Pontus, with physical peril.

    Like that of Dumas's Count of Monte Cristo, Odysseus' journey back to Ithaka, to all that was once his own, uncontested—wife, son, and kingdom—cannot be taken for granted.[18] Nor, apparently, can these be restored without extreme, seaborne suffering. His re-investiture seems instead, first, to require involuntary divestiture:

unforgiving *áskēsis* in the form of a successive series of marine trans-
formations, each with its own steep price and harsh *anástasis*. Not
in the later Platonic sense of death as a kind of purification into
transcendence—a cure for the moral horrors of the soul's prison in
bodily life—but, rather, in the brutal, cleansing deprivations of a
pilgrim, one whose goal is a sacred place, Odysseus is broken down
by the two great waters, Okeanos and Pontus. Ultimately, these
oceans also deliver him home.

Ritual Impurity: Míasma

As well as being a dangerous, mutable, and numinous fact of life,
the ancient Greek sea had unsurpassed cathartic powers, realized not
only metaphorically in epic poetry but also literally in practiced religion. This may have been true from earliest Greek history. Minoan
archaeologist Christos Boulitis suggests that some of the myriad cult
vessels from Cretan palaces, decorated with marine life from octopi
to dolphins, "were most probably also used for libations of sea water,
believed to have cathartic qualities throughout antiquity which is
why it was frequently used in various purification rituals."[19] In the
Greek schema of pollution, water is ordinarily contaminated by the
dirt it washes away, except in the case of the sea, where the opposite
occurs. The sea renders harmless that with which it comes into contact. As the classical text *Etymologicum magnum* says, "Sea water is
purifying by nature."[20]

In ritual cleaning, the sea serves as the most important place to
dispose of pollution and substances that can absorb pollution. Theophrastus's fourth-century B.C.E. satire finds the *deisidaímōn*, the "ex-
cessively holy" or god-fearing "superstitious man," sprinkling himself at the sea.[21] But one encounters this idea as early as the *Iliad*,
when Achilles tells his Myrmidons, racked by the plague of the angry
Apollo as punishment for the theft of the daughter of the god's priest
Chryses, to "wash off their defilement. And they washed it away and

threw the washings into the salt sea" (Οἳ δ' ἀπελυμαίνοντο καὶ εἰς ἅλα λύματ' ἔβαλλον) (*Iliad* 1.313–314). A hecatomb of bulls and goats on the shores of the same "barren salt sea" immediately follows; there can be no acceptable sacrifice to the gods without such purification first. Centuries later, the ancient treatise *On the Sacred Disease* prescribes the disposal of "offscourings" that sponge up evil from purified persons (*kathármata*, the substances per se, such as wool fillets or olive leaves or water, plus the pollution itself), and "they bury some of them in the ground, they throw some in the sea, and others they carry off to the mountains where nobody can see or tread on them."[22]

The ancient Greek understanding of ritual purity featured the sea's unequaled capacity to wash away *míasma* ("stain" or "defilement"), not only a moral condition but also a physical contagion that could contaminate other people. This had to do with a graded scale of purification, whose levels corresponded to the motion of water. In archaic and classical Greece, because in its rushing flow it could carry away *kathármata*, water was the most natural, most basic, and most widely used means of purification. Witness, for example, the ubiquity of the *perirrhantḗrion* (lustral basin) at the entrance to temples. Water had to be pure and thus was most often sought at special springs,[23] as at the famous Castalian springs at Delphi. More important, it could not be stagnant but had to come from a flowing source.[24] In Aeschylus' *Eumenides*, Orestes seeks refuge at the statue of Athena in Athens, insisting on his ritual purity, even though he has murdered his mother Clytemnaestra: "Long since, at the homes of others, I have been absolved thus, both by *running waters* and by victims slain."[25]

In ritual use, water is pure; running water is purer. But as the classicist Robert Parker notes in his book *Miasma,* "The most prized cathartic water was that of the salt-stained sea."[26] In addition to the sea's constant motion, which made it a paradigm of flowing efficacy, seawater's saltiness enhanced its power to purify a thousandfold. Salt was sometimes even added to water to be used in ancient Greek religious ceremonies, perhaps in an attempt to "recreate" seawater.[27]

To purify a holy object or a prospective initiate into holy mysteries—to purify something or someone in anticipation of some *teletaí*, or to make a new beginning—Greeks turned seaward. Before entering the healing Asklepieion, the diseased supplicant had to bathe in the sea, if possible. Before participation in the Great Mysteries at Eleusis, the famous rites of Demeter and Persephone with their strong overtones of fertility and immortality, each *mýstes* (initiate) was required on the second day of the festival in the month of Boedromion to bathe his or her entire body in the sea, along with an individually purchased sacrifice, a "mystic pig."[28] Cult statues were often bathed in the sea to purify them. In a ritual resembling the physical tendance of Hindu divine images, the garments of the all-important statue of Athena Polias on the Athenian Acropolis were purified by washing them in the sea during the annual festival of the Plynteria.[29]

As well as preparatory cleansings, the sea was used as a purifier in the expiation of crimes or in cases of inadvertent pollution. In this respect as well, cult statues were subject to the same ritual exigencies as were people, as becomes clear in a key plot twist in *Iphigenia in Tauris*, discussed later. An inscription from Kos from the first part of the third century B.C.E. stipulates that if a sanctuary is polluted by a dead body, the cult statue must be purified *epì thálasson*, in the sea.[30] The Palladion at Phaleron was carried by Ephebes in an annual procession to the sea. There it was reinstalled in the law court, where those convicted had to retrace the statue's route, sharing in its coastal path of purification.[31] An extraordinary conflation of many of these themes can be found in Pausanias' account of the antagonism between a man and a statue of his dead rival, the local Thasian cult-hero and pan-Hellenic athlete Theagenes. As was so often the case in ancient Greece, the *míasma* of the murderous struggle could be absorbed only by the sea.

> When he [Theagenes] departed this life, one of those who were his enemies while he lived came every night to the statue of Theagenes and flogged the bronze as though he were ill-treating Theagenes

himself. The statue put an end to the outrage by falling on him, but the sons of the dead man prosecuted the statue for murder. So the Thasians dropped the statue to the bottom of the sea, adopting the principle of Draco, who, when he framed for the Athenians laws to deal with homicide, inflicted banishment even on lifeless things, should one of them fall and kill a man. But in course of time, when the earth yielded no crop to the Thasians, they sent envoys to Delphi, and the god instructed them to receive back the exiles. At this command, they received them back, but their restoration brought no remedy of the famine. So for the second time they went to the Pythian priestess saying that although they had obeyed her instructions the wrath of the gods still abode with them. Whereupon the Pythian priestess replied to them: —But you have forgotten your great Theagenes. And when they could not think of a contrivance to recover the statue of Theagenes, fishermen, they say, after putting out to sea for a catch of fish caught the statue in their net and brought it back to land. The Thasians set it up in its original position, and are wont to sacrifice to him as to a god. There are many other places that I know of, both among Greeks and among barbarians, where images of Theagenes have been set up, who cures diseases and receives honours from the natives.

Description of Greece, vol. 6, 11.6–9 (Elis)[32]

In addition to showing the modern reader both the iconic autonomy accorded to statues in ancient Greece and their interchangeability with their subjects, the story is a classic account of the judicial and ritual relief offered by the sea in popular religion. The statue of Theagenes, which probably fell after a particularly energetic flogging, was sued by the kinsmen of the dead man, prosecuted for murder, and punished—and the *polis* was cleansed of its collective tension as Girard might have it, by being "dropped to the bottom of the sea." καταποντοῦσι τὴν εἰκόνα means, literally, they "plunged" or "threw down" the image into the sea; in the case of a human being, the verb can mean "drowned." At least one prominent translation renders it exactly this way, highlighting the ancient Greek oscillation between person and plastic representation, in which statues were essentially animate and "stood for" their subjects in every respect: "They drowned Theagenes' statue in the sea."[33]

Yet the story's *sequentiae* reveal that marine purification was not the final resolution; the sea was compelled through an oracle to reverse its purifying role and surrender what it had swallowed. This was possible, even mandated by Delphi, because Theagenes' posthumous status as a divinized hero of Thasos, with his corresponding powers of fertility and healing, superseded his own statue's accountability for murder. His absence at the bottom of the sea thus created a famine, an abrogation of the fertility of the land of which he was a supernatural guardian.

The sea swallowed "evils," to be sure, thus neutralizing them. In an institutionalized parallel to the Theagenes story, the axe that was found guilty of the slaughter of the ox at the ancient rite of Zeus, the Attic Bouphonia, was thrown into the sea to be rid of it forever. But the sea could do more; it could actually *undo* the worst of offenses. When, in ritual situations, purity from sin's stain (*hamartía*) or the more potent blood-guiltiness (the worst form of *míasma*) had to be absolutely guaranteed, the sea was the best and only hope. Hephaistos cleansed Pelops at the ocean from the blood of Myrtilus, the murdered charioteer of Oenomaos. That which was originally impure, even as desperately impure as a matricide like Orestes, the sea could make pure again: not just pure enough for reintegration into normal social intercourse, but pure enough to be offered to the gods. That is the premise of Iphigenia's claim to the dubious Thoas. The ritual sequence in Euripides' *Iphigenia in Tauris* recapitulates the interlocking sacrificial actions first laid out in *Iliad* 1: purification by seawater must precede immolation.

The Testimony of Tragedy

In three classical tragedies, the cathartic role of the sea in ancient Greek ritual is notably highlighted: Sophocles' *Oedipus the King*, his

Ajax, and finally the *Iphigenia in Tauris* of Euripides. The complex matter of ritual as a topos in Attic drama has been treated extensively in recent years.[34] The dramatic enactment, within theater performances, of rituals and ritual ideas that the Athenian audience would have intimately experienced in public or private religious life, may have been a strategic move on the part of the dramaturges. Its effect may have been to subtly move the audience from their roles as passive spectators of old myths to involved participants, witnesses to stylized "rituals" on stage, making them, as the play unfolded, a kind of meta-chorus. At any rate, it is doubtful that rituals presented in plays, drawing as they did on the deeper wellsprings of lived religious experience, remained confined by their generic containers as mere entertainment for the audience.

Deadly pollution on great scale and its shocking source are the central themes of Sophocles' *Oedipus the King*. The action begins as a plague ravages Thebes, with the antistrophe of the *párados*, the chorus's entrance song, describing the horror:

...in the unnumbered deaths
of its people the city dies;
those children that are born lie dead on the naked earth
unpitied, spreading contagion of death; and grey haired mothers
    and wives
everywhere stand at the altar's edge, suppliant, moaning; the hymn
    to the healing God rings out but with it the wailing voices are
    blended.

Oedipus the King 179–186[35]

The chorus longs to send the war god, whom they blame for the contagion that is destroying them, into the sea. They pray:

There is no clash of brazen shields but our fight is with the War God, a War God ringed with the cries of men, a savage God who burns us; grant that he turn in racing course backwards out of our

country's bounds to the great palace of Amphitrite or where the
waves of the Thracian sea deny the stranger safe anchorage.

Oedipus the King 190–197

Ares ought to be expelled not just "out of our country's bounds"
but "to the great palace of Amphitrite," the sea goddess who lives
beneath the waves, in other words, into the alien sea that alone can
contain and dilute the poison infecting the community. Here the
strength of the chorus's recommendation is drawn from ritual: the
rites of absorptive purification by water, and especially by the sea,
also carry with them the notion of sending or carrying away, typified
by terms such as *apopompé*, "expulsion."[36]

In the ceremony of *apopompé*, "the purifier would emphasize separation from the *kátharma* by 'throwing them over his
shoulder,' and 'walking without looking back.' "[37] As Parker comments upon the passage just cited from *Oedipus the King*, "One
can send away evil persons and evil things to distant regions just
by using words."[38] Thus *apopompé* can be a verbal riddance, a ritualized exile of what is causing trouble. In the view of the chorus,
invoking the theme we noted in chapter 2, the sea is not the inhabited space and so cannot inspire the same protective instinct as
the stable, root-laced land on which a people builds its dwellings
(*oíkoi*), where they worship their gods, and bury their dead (who
also live in *oíkoi*, as this is also the word for "tombs," the houses
of the dead).[39] The land is what is threatened by the deadly pollution, whereas the sea can absorb it, and so seaward is where the
scourge is imprecated to flee. Later, as King Oedipus decries the
miásmatos (the "pollution bearer," 241), the one who has killed
Laius, the son of Labdacos, he urges that the unknown man be
expelled from the *oíkoi* of Thebes.

Of course the plot's dénouement—that the murderer and scourge
is Oedipus himself, that Laius was his father, and that his own wife
also his mother—results in the pathos of the king's self-blinding and
self-exile. By pursuing the truth, Oedipus effects his own *apopompé:*

he is, paradoxically, at once the purifier, the purified, and the abandoned *kathármata*. What is important for our purposes is that the first recourse for the polis is the sea. It is to the sea that first collective impulse for relief is directed. The sea is asked to ritually absorb the civic blight, hideously mirroring in contemporaneous literary images the historical plague that had ravaged Athens only a few years before the composition of the tragedy in 426 B.C.E. The playwright himself survived the plague, provoking him, or so tradition has it, to introduce the healing god Asklepios to the Athenian Acropolis in 425, at essentially the same time that he was believed to have composed *Oedipus the King*.

In Sophocles' *Ajax*, the hero is discovered at Troy after his defeat in the contest for the armor of Achilles. Temporarily insane, he mistakes his own livestock for the rival Greek chieftains and hacks them apart in a bloody frenzy. As Odysseus reports to Athena,

> We found not long ago
> Our flocks and herds of captured beasts all ruined
> And struck with havoc by some butchering hand.
> Their guards were slaughtered with them. Everyone
> Puts the blame on Ajax. . . . How can these prints be his?"
>
> *Ajax* 25–27; 33[40]

When the hero's mind clears, he is discovered at the play's outset ruined by shame, surrounded in his tent by a pile of "victims, slain with his own hand, deep in blood" (219), unable to accept what he has done. He asks,

> How could I be so cursed?
> To let those precious villains out of my hand,
> And fall on goats and cattle,
> On crumpled horns and splendid flocks,
> Shedding their dark blood!
>
> *Ajax* 372–376

But in the next scene, Ajax is a changed man, full of the eerie calm of the suicidal. The agony of intolerable emotion is over, and there is a plan. He has resolved to impale himself on his own sword in the sand of the ocean shore, a place that he represents to his loved ones as the only "cure" possible for the shame of his pollution: "No good physician quavers incantations / When the malady he's treating needs the knife" (*Ajax* 582–583). Bidding good-bye (directly) to his small child, Eurysaces, who presumably cannot understand his words, and (indirectly) to his wife, Tecmessa, Ajax announces his intention to "cleanse himself" by the sea: a plan quite appropriate and indeed even mandatory in the context of ancient Greek ideas of ritual pollution. Only the sea could cleanse the guilty of a slaughter so monstrous, *hamartía*, resulting in a *míasma* so crippling. In lines 654 through 660, Ajax delivers a doomed soliloquy, steeped in irony:

> But now I'm going to the bathing place
> And meadows by the sea, to cleanse my stains,
> In hope the goddess' wrath may pass from me.
> And when I've found a place that's quite deserted,
> I'll dig in the ground, and hide this sword of mine,
> Hatefulest of weapons, out of sight. May Darkness
> And Hades, God of Death, hold it in their safe-keeping.
>
> *Ajax* 654–660

The "stains" or "defilements" (*lúmatha*) to which Ajax refers represent both the animal blood still reddening his hands (10) and the religious pollution that his act has brought upon him. This is the same word used in Achilles' instruction to the Myrmidons (in *Iliad* I.314) and carries the same double valence of both actual filth and moral depravity. Ominous double meanings permeate the passage, all of them steeped in marine ritual. His "digging in the sand" will hide his sword out of sight, but with its hilt anchored and its lethal blade buried in Ajax's entrails. Then it will be again hidden forever, buried with him, as he has requested. And the ritual bathing-place at the shore (*loutrá*) will indeed purify him—but as a landscape for his self-inflicted death. So the sea will cleanse Ajax of the triple pollutions of actual blood, of his blood guilt (*míasma*), and, perhaps, on a metaphorical plane, anticipating the metaphysics of the *Phaedo*, of the stain and burden of life itself.

The last example of sea ritual in tragedy I will consider is the one performed by Iphigenia, the daughter of Agamemnon. The play is Euripides' *Iphigenia in Tauris*, believed to have been written around 414 B.C.E. The plot, the playwright's idiosyncratic version of the myth, is contorted and weird, shot through with head-snapping reversals. The premise, implied in the later *Iphigenia at Aulis*, is that Artemis, dea ex machina, has rescued Iphigenia from the altar at Aulis where she was to be offered as a human sacrifice by Agamemnon and the priest Colchis. The princess of Mycenae was spirited away to the sanctuary of her savior Artemis in the land of the Taurians (modern Crimea). There, by the "barbarian shore," in the dark paradox of her legend, she herself must consecrate the victims of a human sacrificial cult to Artemis: Greek strangers who land at the shores of Tauris. The ceremony that Iphigenia must perform involves sprinkling each victim with the water of the sea (*hudraínein*, e.g., *Iphigenia in Tauris* 55),[41] thereby rendering him pure and acceptable as an offering.

As fate would have it, it is her brother Orestes himself who, having sailed through the Clashing Rocks, lands "by the dark sea-wash" in Tauris. Despite the Athenian tribunal's absolution of his murder of their mother Clytemnaestra, he is still pursued by the avenging Erinyes. At Apollo's command, he seeks in Tauris the carved, "heaven-fallen" cult image of Artemis, whose installation in Attica will acquit him of the last stain of his matricide. He and his companion Pylades are discovered, captured, and taken to the Taurian king,

Thoas, by herdsmen who have gone to the shore to wash their cattle in sea brine. Paralleling the madness of Ajax, Orestes, maddened by the Furies, had sliced into the animals with his sword, thinking they were his ancient snaky tormenters. Iphigenia is ordered to prepare the foreigners for ritual slaughter.

Through a series of intensive dialogues, poignantly revolving around Orestes' desire for his lost sister to prepare his body for burial rites, brother and sister are at last revealed to one another. Iphigenia is desperate to be free of her grim indenture. But drawing on her experience as a priestess of human sacrifice, she will exploit the collective belief in the purifying power of the sea's waters. Her escape plan to divert king Thoas involves forms that she knows will be intelligible to him: ritually cleansing in seawater both Orestes and the Artemis statue, which she will claim her brother has touched, contaminating the goddess, who now turns her head away. Iphigenia can succeed only by being adamant about this: "I wish to sanctify you [*hagnízein*, literally, "make you *hagnós*, "holy"][42] with the springs of the salt sea" (πόντου σε πηγνὶς ἀγνίσαι βουλήσομαι, *Iphigenia in Tauris* 1039). When, to Thoas' horror, it emerges that the strangers are fouled with the blood of murdered kin, she also offers the remedy:

> THOAS: *How shall we deal with the strangers?*
> IPHIGENIA: *We must respect the law of the sanctuary.*
> THOAS: *Your pitcher and great knife, then, for the sacrifice?*
> IPHIGENIA: *First I want to immerse them, for the sake of holy purifications.*
> THOAS: *In the waters of springs, or in the salty sea?*
> IPHIGENIA: *The sea can wash away all evils of humankind.*
> THOAS: *It might be more sacred, then, to the goddess.*
> IPHIGENIA: *I myself think it is better.*
>
> *Iphigenia in Tauris* 1188–1195[43]

θάλασσα κλύζει πάντα τἀνθρώπων κακά (the sea can wash away all evils of humankind), that is, that are incurred by, belong to, or even "adhere" to humankind (*Iphigenia in Tauris* 1193), in that sticky way *míasma* has of hanging on to the very flesh, and most certainly of pervading one's destiny. All these senses are implied by the genitive plural, *ánthrōpōn*, for all are included in the Greek concept of "evils." No stronger or more revealing assertion of the sea's cathartic powers exists in ancient Greek literature. Ironically, Iphigenia says this not because she will in fact purify Artemis and the Argive fugitives, but because Thoas thinks she will, and because he is convinced that it will work. The sea will de-contaminate Orestes and Pylades so thoroughly that even they can be worthy sacrificial victims to propitiate an offended goddess.

This is a plan that will use the rituals of the sea as an elaborate ruse. But it will also, in truth, use the sea as its vehicle of salvation and escape. Iphigenia says to Orestes,

> Yes, risk the sea.
> You challenged it, came through it. Having once
> Met it and mastered it, you can again.
> And so let fly your oars. Yes, risk the sea,
> Take to the ship—though who can surely tell
> If God or man shall steer you through the waves
> To a safe landing, or if Fate shall grant
> Argos the benison of your return?
> Or me—who knows?—the sweet surprise of mine.
>
> *Iphigenia in Tauris* 893–897

The play climaxes with the dark inevitability and solemnity of an impending human sacrifice as the priestess of Artemis, Iphigenia, emerges from the temple doors to lead the procession down the steps to the sea. To maintain the trick, she cries out to the crowd, warning of the contagion brought by Orestes and Pylades, but again clearly speaking in a believable religious idiom about their alleged condition:

> Taurians, turn away from the pollution.
> Gate-tenders, open the gates, then wash your hands.

Men who want wives, women who want children,
Avoid contagion, keep away, keep away!

Iphigenia in Tauris 1226–1229

She lifts the contaminated statue of the goddess high:

"Oh Virgin goddess, if the waves can wash
And purge the taint from these two murderers
And wash from Thee the tarnishing of blood,
Thy dwelling shall be clean and we be blest!
To Thee and the all-Wise my silent prayer."

Iphigenia in Tauris 1230–1234

Under the pretense of wading far into the sea to conduct the cleansing rites, all three Greeks escape in the ship concealed nearby: the fugitive Orestes, his friend Pylades, and his sister Iphigenia, bearing the statue of Artemis, "the image of the Daughter of High Zeus" (1385).

We heard a glad voice ringing through the ship,
"O mariners of Hellas, grip your oars
And clip the sea to foam! O let your arms
Be strong, for we have won, have won, have won
What we set out to win! Soon we shall leave
The jagged Clashing Rocks behind! Pull hard!"

Iphigenia in Tauris 1386–1389

Thoas' anger at the report of the escape is inflamed by the ship's entrapment against a treacherous Taurian cliff. He is determined to capture and kill them all, when Pallas Athena appears and tells Thoas that she has ordered Poseidon to calm the sea and that the ship is on its way to Attica. She has ordered Orestes to build a temple to Artemis at Halae, on the outskirts of Attica, and to set up the image of Artemis there.

And let this be the law. When they observe
Her festival, the priest shall hold, in memory of you, the sharp
   blade of his knife
Against a human throat and draw one drop
Of blood, then stop—this in no disrespect
But a grave reminder of her former ways.

Iphigenia in Tauris 1456–1462

Iphigenia is charged with keeping the keys at Artemis's shrine at Brauron:

And at your death you shall be buried there
And honored in your tomb with spotless gifts,
Garments unworn, woven by hands of women
Who honorably died in giving birth.

Iphigenia in Tauris 1463–1467

Artemis, ever virgin, in the paradoxical duality so characteristic of ancient Greek godhead, is the patronness of childbirth, and such sad offerings thus fill her windy temple treasuries.

When one January I visited Iphigenia's tomb at Brauron, out in the rushy marshes, and saw it surrounded by the archaic graves of the sanctuary's priestesses, I thought of how intertwined her story was with certain themes: with the virgin huntress Artemis; with that goddess's bloody love of human sacrifice, both executed and averted; and finally, of course, with the sea. Tricked into coming as Achilles' bride to the shore at Aulis, Iphigenia, wearing flowers, is led like a heifer to the altar to be slain, a bloody wind-charm to persuade Artemis to release the becalmed Achaean fleet. Her throat is cut (in some accounts) by her own father, or else by a priest, with Agamemnon's permission. In the "rescued" version of the myth, she is whisked over the sea past the Clashing Rocks to preside by a foreign shore, in the service of Artemis, over human sacrifices herself; and finally, using the sea's cleansing powers as her own trick, she escapes the Taurians by sailing with her brother back to Greece. There, however, Iphigenia does not elude her double helix spiral with the goddess. And I thought as I looked down at her stone-lined grave, even though it was an archaic-period cenotaph—at least 650 years too late, perhaps more, for the aftermath of the Trojan War, and empty as well—how strange it was that she should lie here, even in myth, so far from the sea.

Then my friends called me to walk with them a few feet up the road and over a tree-covered hill. At the foot of the hill stretched the sand, and beyond the sand crashed the waves. I had not looked closely at the map before we left Athens. Brauron, too, is by the sea. Iphigenia is home, all evils washed away.

CHAPTER 5

"The Great Woman Down There"

Sedna and Ritual Pollution in Inuit Seascapes

All our customs come from life and turn towards life; we explain nothing, we believe nothing. . . . We fear the weather spirit of earth, that we must fight against to wrest our food from land and sea. We fear Sila. We fear dearth and hunger in the cold snow huts. We fear Takanakapsulak, the great woman down at the bottom of the sea, [who] rules over all the beasts of the sea.

Aua, Iglulik Inuit shaman

On November 14, 2003, three astronomers at Mount Palomar Observatory, led by California Institute of Technology's Michael Brown, discovered a small red planetoid of rock and ice, deep red in color, currently eight billion miles from the sun—three times farther away than Pluto and three-quarters its size. The object's elliptical orbit is unlike that of any other planet and takes 10,500 years to complete, leading it as far out as eighty-four billion miles from the sun. Within hours after the announcement of its discovery on March 15 of the following year, and before scientific debates about what constitutes a "planet" culminated in 2006, the body was heralded as a possible tenth planet. Deep in the outmost lying reaches of the solar system, with a surface temperature of about −240° C (−400° F), Celestial Object 2003 VB12 was named by its discoverers after the volatile Inuit spirit who dwells at the bottom of the frigid ocean: Sedna, "the great woman down there."[1] Deep, dark space becomes an analogue to the depths of the lightless sea, and the remotest denizen of space becomes the mercurial, loveless, bottom-dweller of the sea known to the circumpolar peoples of North America and Greenland. In naming the new object Sedna, "up there" becomes "down there" in both the scientific and the wider cultural imagination, for both realms—space and sea—are cold, lost, and utterly distant.

This is an exploration of an ocean-centered ritual complex of social reparation, one that, contrary to much that is assumed about the processes of modernization and cultural deterioration, is still very much alive in the Inuit imagination, if the intensity of artistic expression is any indication.[2] The themes of this complex are the familiar ones of pollution, repentance, and purification at sea. Their vehicle, however, is the wrath and appeasement of a goddess on the ocean floor, signified in a startling, graphic way: human transgressions enrage Sedna, the great Sea Woman, and make her beautiful black hair wild and disheveled, crawling with lice and full of debris. Taboo violations plug her mouth and eyes with dirt and painful irritations. In her wrath, the goddess withholds the sea animals of the hunt, and the people starve. Only an entranced shaman, an *angakoq* (or, to the Iglulik, *nakazoq*, "one who drops down to the bottom of the sea") can make Sedna's hair and face clean again, as he aggressively questions her about what has caused her condition. Only the *angakoq* can rebraid her hair, since Sedna's fingers were hacked off by her own father, and only he can return along his dangerous path back up into the air.[3] A bargain is struck: repentance in exchange for the release of the beasts of the sea. A public confirmation of these transgressions, exacted in fear back in the ceremonial house by an angry shaman upon his return, catalyzes moral restitution. The seals, walruses, and whales swim up from their deep confinement to ride the waves once more and to feed beneath the ice.

I have already discussed Mary Douglas's insistence on the convergence of systems of ritual purity with those of moral and ethical codes, and her equally strong assertion that this ubiquitous polarity is a cultural vehicle, having virtually nothing to do with literal contamination—that "dirt" is a state of mind. However, building upon the critique of sociologist Kevin Hetherington referenced in chapter 1, I implied that to view "dirt" as an arbitrarily assigned, purely cultural category ignores the genuine dirtiness of dirt. In the legend of Sedna and its ritual reenactments, the sea's cathartic cycles are not only divinized, but also urgently somatized. Sea Woman physically embodies the imperatives both of marine pollution and of marine purification, as well as their strong contingency upon variable human ethical behavior. Because of ritual violations, her hair is literally dirty, and thus she suffers. The dirt is both emblem and cause of her pain; that pain is both the physical discomfort of being unkempt and the psychic agony of having been betrayed. Its cleansing requires moral but also very physical intervention by a ritual expert, who, in Inuit ceremonial praxis, alarmingly combines the roles of intercessor and hairdresser.

The Myth of Sedna

Sedna was once a human girl who refused any husband and so was married by her father to a fulmar, a sea bird—or, in other versions, a dog—and who, after a series of horrifying twists of fate, now dwells at the bottom of the sea, where she is the mother of all sea mammals. Like Iphigenia, a Greek princess turned high priestess of human sacrifice in a foreign land, the Inuit deity Sedna is assaulted by male hands, her own father's. Once a human woman of social rank in her own land but becoming a displaced exile, she is ultimately a sacrificial victim at sea. "Sedna is a blood sacrifice and the sea is the altar on which this sacrifice takes place."[4] After surviving arbitrary violence, also like Iphigenia, Sedna then undergoes yet another metamorphosis, this time from a human to a supernatural figure, and is invested with uncanny powers. Like Iphigenia, who is physically "translated" from Aulis to Tauris at the moment of her slaughter to become a hierophant of Artemis, Sedna controls the sea in sacerdotal fashion, determining its relationship to human beings.

The difference between the two narratives is that Sedna serves no goddess but instead becomes one or, more accurately in Inuit theological idiom, becomes an *inua* in her own right[5]—and an ambiguous and capricious one, wielding powers of life and death as they were so horribly wielded against her. The powers that she exercises encompass sea creatures, the animals of the Inuit hunt, who once were her own fingers. Severed fingers, filthy hair, clogged eyes and mouth, and cold exile in the watery abyss: Sedna both instigates and responds, in her own body, to the cycles of human moral pollution and its ritual redress through repentance and purification.

This redress, which comprises both physical and moral cleansing, leads to the release of her children: seals, walruses, and whales. Although the paradigmatic drama plays itself out on both land and sea, it is Sedna's own body that is the central theater of action. In the logic of this performative sequence, that divine body is, I will argue, a microcosm and homologue of the sea itself and therefore endures its highly ethicized cycles of pollution and purification. Even though the sea preexists Sedna in her mythology, Sedna becomes its animating spirit by means of her cruel, involuntary plunge into it. The sea is the matrix and source of the animals, but Sedna is the *inua* of the sea: Saittuma Uva, "Spirit of the Sea Depths," as she is called in Greenland. The sea animals are thus of and from her, as she is the sea's *inua*, without which it has no agency and no generativity.

Sedna does not "symbolize" the sea. Instead, the Inuit concept of *inue*—the indwelling spiritual forces ("breath-souls") of both humans and animals—extends to natural elements: Wind Indweller, Sea Mother, Moon Man. But as Daniel Merkur warns, *inue* are not abstract entities, nor are they exactly "personifications of nature," as Rasmussen described them.[6] Instead they are personal superpowers, who animate and are identified with their particular environments: "ordinarily invisible forces immanent in physical phenomena."[7] *Inue* are ideational in both content and function: "The Labrador *inua* is 'a thinking spirit,' and the West Greenland *inua* is the spirit of anything that 'can be said to form a separate idea.'"[8]

Western philosophic analogies fail to be precise . . . because *inue* are not impersonal ideas but personal beings. An *inua* is, at bottom, an idea that indwells in and imparts individual character to a phenomenon. As such, it has, employs, and most essentially is a power . . . *inue* are present within the phenomenon and imparts form to it even as it differs from the phenomenon's substance. Because it is both an idea and has active power to implement the idea in substance, an *inua* is actively engaged in thinking. *Inue* thought is both structural, imparting the ideas that they are to the phenomena in which they indwell, and human-like or anthropopsychic. They think by means of verbal ideas. They have emotions and motivations, and they can communicate with Inuit. Because *inue* are anthropopsychic, they are personal beings and may be personified.[9]

This metaphysical reality requires the Inuit to live in nuanced relationship to the natural world, entailing both interpersonal and theological complexity. The result of such a thought construction is that Sedna's personality extends to the perimeters of her realm and the sea behaves as she does, reflecting her tormented past: "Because the Sea Mother is jealous and vindictive, the sea is dangerous and miserly in its provision of game."[10]

The mythic cycles of Sedna, called by the Iglulik *Takanaluk Arnaluk*, "the great woman down there," and the reports of her annual autumnal conquest by shamans, are best known from studies of the Baffin Island Inuit in the late nineteenth century by Danish ethnologist Franz Boas, but are amplified by, among others, the concurrent ethnographies of Knud Rasmussen and the later studies by Diamond Jenness, Erik Holtved, Inge Kleivan, Åke Hultkrantz, and, most recently, Daniel Merkur.[11] Her legend is known among the Inuit over an area reaching as far as southwestern Alaska, and she also is known by other regional names, among them the polar Inuit's Nerrivik, "Food Dish," the west Greenland people's Sasvsuma Inua, "Indweller of the Deep," the Netsilik's Nuliajuk, "Lubricious One," the Iglulik's Takanakapsaluk, "The Terrible One Down There," and, as though by way of explanation of her fate, the Baffin Land Inuit's Uinigumissuitung, "She Who Never Wished to Marry."[12]

Myriad versions of the Sedna myth exist; all of them, however, tell a tale of mutilation and loss: pride, lies, cruel reversals, social ostracism, catastrophic fear, vengeance, and, ultimately, uncanny change, in the form of an irreversible descent from a terrestrial life to a submarine one. In most, although not all, versions (the Netsilik tale, for example, is about an orphan girl cast overboard by her people),[13] the underlying theme includes the fracture of the ultimate taboo, which mandates that human beings must not marry animals. The *telos* of Sedna's story is a wild transformation, accompanied by and the ironic restoration of new supernatural powers.[14] In *The Inuit Imagination: Arctic Myth and Sculpture,* Harold Seidelman and James Turner offer a synopsis of the story, the "fulmar" version, first reported to white Europeans and Americans by Boas.

Once upon a time there lived on a solitary shore an Inung with his daughter Sedna. His wife had been dead for some time and the two led a quiet life. Sedna grew up to be a handsome girl and the youths came from all around to sue for her hand, but none of them could touch her proud heart. Finally, at the breaking up of the ice in the spring a fulmar flew over the ice and wooed Sedna with enticing song. "Come to me," it said; "come into the land of the birds, where there is never hunger, where my tent is made of the most beautiful skins. You shall rest on soft bearskins. My fellows, the fulmars, shall bring you all your heart may desire; their feathers shall clothe you; your lamp shall always be filled with oil, your pot with meat." Sedna could not long resist such wooing and they went altogether over the vast sea. When at last they reached the country of the fulmar, after a long and hard journey, Sedna discovered that her spouse had shamefully deceived her. Her new home was not built of beautiful pelts, but was covered with wretched fishskins, full of holes, that gave free entrance to wind and snow. Instead of soft reindeer skins her bed was made of hard walrus hides and she had to live on miserable fish, which the birds brought her. Too soon she discovered that she had thrown away her opportunities when in her foolish pride she had rejected the Inuit youth. In her woe she sang: "Aja. O father, if you knew how wretched I am you would come to me and we would hurry away in your boat over the waters. The birds look unkindly upon me the stranger; cold winds roar about my bed; they give me but miserable food. O come and take me back home. Aja."

When a year had passed and the sea was again stirred by warmer winds, the father left his country to visit Sedna. His daughter greeted him joyfully and besought him to take her back home. The father hearing of the outrages wrought upon his daughter determined upon revenge. He killed the fulmar, took Sedna into his boat, and they quickly left the country which had brought so much sorrow to Sedna. When the other fulmars came home and found their companion dead and his wife gone, they all flew away in search of the fugitives. They were very sad over the death of their poor murdered comrade and continue to mourn and cry until this day.

Having flown a short distance they discerned the boat and stirred up a heavy storm. The sea rose in immense waves that threatened the pair with destruction. In this mortal peril the father determined to offer Sedna to the birds and flung her overboard. She clung to the edge of the boat with a death grip. The cruel father then took a knife and cut off the first joints of her fingers. Falling into the sea they were transformed into whales, the nails turning into whalebone. Sedna holding on to the boat more tightly, the second finger joints fell under the sharp knife and swam away as seals; when the father cut off the stumps of her fingers they became ground [bearded] seals. Meantime the storm subsided, for the fulmars thought Sedna was drowned. The father then allowed her to come into the boat again. But from that time she cherished a deadly hatred against him and swore bitter revenge. After they got ashore, she called her dogs and let them gnaw off the feet and hands of her father while he was asleep. Upon this he cursed himself, his daughter, and the dogs which had maimed him; whereupon the earth opened and swallowed the hut, the father, the daughter, and the dogs. They have since lived in the land of Adlivun, of which Sedna is the mistress.[15]

Dismembered, alienated from both marital and birth families, and socially crippled forever in human contexts, Sedna dwells at the bottom of the sea in her realm, Adlivun. There she rules as a kind of "Mistress of Animals," borrowing the archaeological name for a similar figure in Bronze Age Asia Minor and Crete, a goddess whose realm includes control of "her children," the marine animals. In the course of her ordeal she has undergone a fearsome metamorphosis. Inuit sculptors to this day depict Sedna as half woman half seal, whale, or fish, with a great powerful tail and a face pregnant with suffering and a great powerful tail. Like Tolkien's Gollum, once the hobbit Sméagol, Sedna's personal history has left scars so deep that they manifest in permanent bodily distortion; she is left only half-human, a creature, a *Mischwesen*. In some stone carvings, she has outrageous hair,[16] and in others she has a thick, neat braid, depending on which phase of her oscillating condition most interests the artist. Not surprisingly, Adlivun is also one of the three destinations of the human dead, where a diseased soul spends a year or more with Sedna in a kind of purgatorial penance before traveling to Omiktu, heaven.[17] Sedna is also the Mistress of the Dead; her role as keeper of the laws spans both human and animal realms, both the living and the dead.

Marine Animals and Taboos in the Sedna Cycle

Although the Sea Woman legend is familiar to nearly all Inuits, the shamanic experience of cultically intervening with Sedna is limited to coastal-dwelling peoples, from the Copper Inuit in western Canada to the Inuit of east Greenland. The Copper Inuit ritual requires the *angokoq* to hook Sea Woman like a great fish and drag her to the surface to subdue her by force, but more common is the shamanic trance, witnessed by the community, that sends him to where she sulks. Far below, in her dark house, Sedna prevents her children, the sea beasts, from rising to the hunt, and they prevent themselves.[18]

The undersea journey ritual has a number of variants, with different understandings of the form of the pollution in Sedna's hair, that makes her uncomfortable and violent. Violations of traditional observances in the charged spheres of hunting, birth, and death are always the genesis of Sedna's affliction.[19] Among the Copper Inuit, such violations may involve excessive sewing by women on ice or breaking cooking taboos; among the east Greenlanders, they are transgressions of animal ceremonialism. In the Cumberland Sound area of Baffin Island, Sedna was known to dislike land animals, especially the caribou, so land creatures and sea creatures could not be hunted or eaten together. On the west coast of Hudson Bay, caribou hunters would place small pieces of sealskin under rocks to appease her.[20] In the same area, the souls of sea mammals whose bodies were maltreated after the hunt or who suffered the breach of death taboos, traveled "back home" to Sedna's house.[21] Such transgressions took the form of irritating attachments to the animals' souls, "representing the injustices they had suffered."[22]

It is difficult to exaggerate the interpenetration between the realms of hunter and hunted in Inuit ethical thought.[23] Consider the remarks of the elder Lucassie Kattuk (b. 1928) of Sanikiluaq, Nanuvut: "I have always hunted for seals. I really enjoy seals. We do not play around with animals. We respect animals. We kill it as quickly as possible. This is how we do it. We appreciate the animals. We hunt for them."[24] The importance of Sedna's motherhood—and control—of the sea animals, simultaneously the subjects and objects of the hunt, emerges readily from such statements, and the ethical contingencies of this control appear inevitable. Human transgression can be adequately expressed and indexed only by the pollution of the sea's spirit, its incarnate female "thought." It results in famine and the categorical impounding of the animals of the hunt. Furthermore, the sea animals are far from Sea Woman's pawns: based on the degree of ritual pollution among their human hunters, they actually withhold themselves as prey. Baleen whales, seals, and walruses, ensouled beings with special powers to discern human uncleanness, decide who may hunt and kill them. They thus act as agents of Sedna's role as the living, supernatural barometer of human scruples—extensions of her thought world, much as angels in monotheistic traditions have agency yet are created ultimately only the thoughts of the one God. In this intricate net of ideas and religious practices, environmental imbalance has moral dimensions.

Since the nineteenth century, ethnographers have noted the far greater complexity of taboo rules (and hence the far greater likelihood of their violation) observed by seaside dwelling, as opposed to inland dwelling, Inuit. Rasmussen believed that this had to do with the volatile nature of the marine environment itself and its unpredictable food supply. In 1930 he wrote,

> The religion of the Caribou Eskimos was a pronouncedly inland religion, essentially different from that of the coast dwellers. . . . Their comparatively few taboo rules and the much more simple birth and death customs, made it clear to us that they lived under natural conditions that traditionally were indigenous and natural to them; if only they could obtain sufficient food they always felt secure in their surroundings. Those Eskimos who, on the other hand, had made their way down to the sea, had there come across something new, something that to them was quite unknown and strange, something that made them feel very insecure and had thus given birth to all their very intricate precautions in the form of taboo rules and in changes, more complicated religions.[25]

The "intricate precautions" were legion, and they tended to revolve around the life cycles of sea animals and the weather, both in Sedna's realms of power. Persuasive and apotropaic practices included (and still include) the wearing of amulets and the singing of traditional songs of reverence before and during hunting or fishing expeditions. Small offerings to the sea goddess and her creatures are recorded from the Netsilik area (*kiverfautit*: small sealskin bags containing harpoon heads, and carvings of seals, resembling the ritual logic of Haitian *voudun*)[26] to Labrador: "broken knives, worn out harpoon heads, pieces of meat and bone" thrown into the waters.

Expiatory practices were more fearsome. In the Cumberland Sound area, Boas noted a wide range of ritual domestic interdictions after a seal's or whale's death, which were to be fastidiously observed. Individual households were forbidden, among other things, from scraping window frost from the igloo, airing bed coverings, removing drippings from oil lamps, removing hair from skins, melting snow for water, and working with iron, wood, stone, or ivory.[27] Structurally, all these forbidden activities seem to reflect human changes in the material environment. In the delicate calculus of metaphysical danger and propitiation brought to bear on the hunter and his family by the act of killing, human agency had to be eradicated.

Personal bodily taboos, which might be expected to be the province of the human community, have instead traditionally fallen within the ethical purview of the sea beasts. It was incumbent on women to announce publicly the arrival of their menstrual periods and to confess having had an abortion, or otherwise to risk offending the sea animals.[28] Boas was told that the animals could see a kind of "vapor"—the ancient Greeks would have recognized it instantly as *míasma*—surrounding a hunter who had been exposed to unclean persons. Out of aversion, the animals would not volunteer to rise to the sea's surface; that is, they would not consent to be hunted by the ritually impure. One thinks of Iphigenia's words: "The pure alone I slay." The Inuit saying, spoken by the seal, might instead be, "The pure alone slay me."

The words of the Iglulik shaman Aua, recorded by Rasmussen and cited at the beginning of this chapter, add an eschatological dimension to Kattuck's expression of respect for the hunted seals and spell out the implications of the Inuit idea that animals possess a human *inua*. After referring to the fear of "the great woman down at the bottom of the sea," Aua makes the link between Sedna's role as Mistress of Animals and the extreme danger that attends the hunt, thereby making mandatory the rituals of propitiation and expiation:

> We fear the souls of dead human beings and of the animals we have killed. *The greatest peril of life lies in the fact that human food consists entirely of souls.* All the creatures that we have to kill and eat, all those that we have to strike down and destroy to make clothes for ourselves, have souls, like we have, souls that do not perish with the body, and which must therefore be propitiated lest they should revenge themselves on us for taking away their bodies.[29]

The violation of laws surrounding the death of hunted animals (such as the failure to pour fresh water into their mouths after the kill, a time when they were thought to be especially thirsty) is retained by the animals' souls. In Sedna's house, dead sea creatures communicate their distress to those yet unborn. These, her future children, will in turn refuse to surface and give up themselves as prey. The regulatory ethical powers of Sedna as the sea and those of her animals are inseparable, for one is a function and expression of the other. In turn, the animals and the Inuit are mutually identified from the beginning of time. The animals occupy the symbolic "swing" position in this web of relationships, for Sedna and human beings are not only not interchangeable; they are also locked in perpetual antagonism.

Shamanic Intervention

Sedna's myth does not stop with her tragic story but permeates the lived Inuit experience insofar as it drives an etiology for hunting cycles and mandates ritual intervention. Because of their close relationships with sea animals, which Inuit genealogies insist once took the form of a shared language with human beings and the mutual capacity for metamorphosis,[30] Sedna and the coastal communities are compelled to interact. This sphere of interaction has ethical as well as ritual dimensions, although its symptoms are highly empirical. When Sedna is angry, she withholds her children; her children withhold themselves; and the people starve. As Merkur contends,

> Neither the basic temperaments of the indwellers nor the conse-quent characteristics of the phenomena in which they indwell are determined by human activity. However, because *inue* are anthro-popsychic, they are not beyond the reach of social intercourse. *Inue* that undergo periodic change may be the objects, variously, of sacri-fice, propitiation, and reconciliation by Inuit intending to persuade them to change in directions beneficial to humanity.[31]

Sedna is the example par excellence of this cultic dynamic. As Boas, Rasmussen, and others have recorded, these cycles of abundance and scarcity are causally interdependent with the buildup—and ab-solution—of community wrongdoings. Sometimes individual pro-pitiation is not enough, and a ritual expert is required. "I must never offend Nuliajuk," one Netsilik hunter said to Rasmussen.

> I must never offend the souls of animals or a *tuunraq* [helping spirit] so that it will strike me with sickness. When hunting or wan-dering inland I must as often as I can make offerings to animals that I hunt, or to dead who can help me. . . . I must observe my forefa-thers' rules of life in hunting customs and taboos, which are nearly all directed against the souls of dead people or dead animals. I must gain special abilities or qualities through amulets, I must try to get hold of magic words or magic songs that either give hunting luck or are protective. If I cannot manage in spite of all these precautions, and suffer want or sickness, I must seek help from the shamans whose mission it is to be the protectors of mankind against all the hidden forces and dangers of life.[32]

When ritual and moral transgressions accumulate in a particular community, Sedna's anger engenders endless storms, and her hair literally streams out from all sides. Or she wreaks starvation by keeping back her children, the sea animals of the hunt, keeping them in her home on the ocean's floor. Sedna must be lured out of the sea through a hole in the ice by the *angakoq*, negotiated with, and wrestled into submission.[33] Using traditional ecstatic media, drumming and chanting, the *angakoq* and the people invoke the spirit helpers, as in this Iglulik song, sung by the elders:

> We reach out our hands
> to help you up;
> we are without food;
> we are without game.
> From the hollow by the entrance
> you shall open,
> you shall bore your way up.
> We are without food,
> And we lay ourselves down
> holding out hands
> to help you up.[34]

An alternative diagnosis for the lack of seals could be discerned during a trance. As with all these encounters, attended by others, the *angakoq*, possessed by Sedna, would speak in the powerful, low voice of the goddess to lay out before the Inuit the offenses committed against her. As Rasmussen reports, this epiphany would, in turn, immediately catalyze the participants' confession of broken taboos. Among most other Inuit groups, however, it is the burden of the shaman to go to her. While nearly all shamans have been male, a few female shamans are known. Only the greatest shamans can survive a direct and intimate encounter with Sedna.

In a trance and in the presence of the gathered community, the shaman must leave his or her body, headfirst, and travel across a great abyss to the bottom of the sea: "The way is made ready for me; the way is opening before me." He must thread a path between three great, rolling stones, whose gap continually shifts and then follow a trodden path like an earthly coastline. At last he arrives at Sedna's underwater house of stone in the middle of great plain, its open roof allowing her to observe the Inuits' deeds and misdeeds.[35] In some versions of the legend, Sedna's world is guarded by her dog husband; in others, her father, damned to spend eternity with the daughter he betrayed and is thus overflowing with bitter eagerness to torture the newcomer. "A wall in front of the entrance displays her hostility to human kind."[36] The shaman breaks down the wall and fearlessly casts aside the dog. "I am flesh and blood!" he calls out to the father, so as not to be confused with a dead soul, and is allowed to pass;[37] but Sedna sits with her back to her lamp and to the constrained animals gathered around her, in pitch darkness.[38]

As a sign of her anger—but also, paradoxically, of a kind of powerlessness, an objective correlative of human wrongdoing—Sedna's long black hair is unkempt and wild, infested with human evils and ritual transgressions in the form of lice:[39] "filthy with the misdeeds of the Inuit." Among the west Greenlanders, it is said that the parasites, *agdlerutit,* a plural word also referring to abortions and stillborn children (the worst of human pollutions affecting Sedna), fasten themselves onto her hair as transgressions accumulate.[40] Her mouth and eyes are clogged with dirt.[41] On the west coast of Hudson Bay, "evil skins" make the goddess's eyes smart.[42] She cannot see. She cannot breathe. Because of her intimate connection with the violation of taboo, Sedna is a victim once more.

What is really in Sedna's hair, and why can human sin dirty it? What is the conceptual logic underlying this powerful image?[43] As for the "heart" of the symbolic convergence, Merkur points to the enduring Inuit belief—attested to first by John Murdoch at Point Barrow in 1892 and later confirmed by Diamond Jenness in 1922 and then by Nicholas Gubser in his 1965 study of the Nunamiut—that malicious ghosts and spirits appear only in fog and darkness.[44] Because the souls of slain animals are pained by violations of observances and because hair contains the "free soul," the essence of a being untethered by a body (again, including that of the never-born), Merkur suggests that "the revenge-seeking ghosts take lodging in the Sea Mother's hair. When the animals' ghosts are fouled and foggy, her hair becomes filthy as well."[45] Sedna's hair is full of dead animals and lost babies—angry ghosts.

Since her father long ago cut off Sedna's fingers, she cannot remove the dirt and parasites or clean and braid her hair herself. As the Yup'ik museum anthropologist Anne Fienup-Riordan observes, "Whatever her name, her maimed hands from which sea mammals originate set her in sharp contrast to the ideal Inuit hunter and his wife who hunt and care for animals with whole, clean hands."[46] In terms of Inuit values of bodily identity, whereby "people are, literally, the sum total of their hands and feet,"[47] power is optimally expressed through the hands and can be restricted by binding the hands and feet. Sedna is marginalized most of all by her fingerlessness.

In Boas's Baffin Land account, Sedna, her fingers gone, exacts revenge on her father by maiming him in return, letting her dogs gnaw off his feet and hands while he sleeps. In the eastern Canadian Arctic, an Iglulik shaman who journeys to the ocean floor to "comb away the impurities that lodge in her [Sedna's] hair, a task she is unable to perform for herself," wears nothing but mittens and boots, an emphasis through the use of ritual clothing on what both Sedna and her father have lost, the all-important extremities that enable work and define the individual.[48] The hands of the shaman, homologized to the flippers of sea mammals, become the means of purification and efficacy; he is a therapeutic substitute for Sedna, who is powerful and dangerous yet, through her tragic history, impaired, and like the sea that she incarnates is unable to meet the basic challenge of cleaning herself.

Sedna's Braid

Taking her shoulder, the *angakoq* first turns Sea Woman's face to the lamp. He must tend the goddess, straightening, cleaning, and combing Sedna's hair.[49] While working, the *angakoq* presses Sedna on what breaches of taboo have sullied her appearance, reproaching her: "Those above can no longer help the seal up out of the sea."[50]

Sedna retorts (for after all she has been watching through her open roof), "It is your own sins and ill doing that bar the way." A lengthy interrogation ensues, after which the sea goddess finally enumerates which particular ritual laws have been violated and made her such a mess. Polar Inuit shamans beautifully braid and tie back her hair— or, in the Copper Inuit version, comb and smooth it. In keeping with the very material expression of the values encoded in this encounter, Sedna's restored braid (or two braids) is not without strong gender-related significance.[51] Inuit woman traditionally wear their hair long, signifying spiritual power and fertility, but in braids to disencumber their daily activities; accordingly, a neat braid is a quintessential female signifier of order.

Some Inuit traditions have the shaman, disguised in walrus tusks, clean Sedna's house—also filthy—as well.[52] Her ritually charged dirt is absorbed by the ocean that is her matrix and habitus. Singing magical songs, the *angakoq* calms Sedna and appeases her anger. Then, on the condition that the people will again·observe her laws, Sedna releases the animals.[53] "She takes the animals one by one and drops them on the floor. A whirlpool arises on the floor. Water pours out of the pool, and the animals disappear out into the sea, indicating rich hunting for the Inuit."[54]

"With promises to rectify the people's inconsiderate behaviour,"[55] the shaman retraces his or her steps to the ceremonial house on the earth's surface and, emerging from trance, delivers a sermon of fire. In fear, the people confess their transgressions and violations aloud:[56] this step of public exposition, interestingly, is essential to the reparative process, even though the shaman has already learned the sordid history from Sedna. The *angakoq* predicts the return of good hunting, and the animals now swim free once more.

Whereas purity is clearly the central discourse in the cultic sequence considered here, as it was in the case of the Attic tragedies that treat the sea as a supreme means of catharsis, there is at the same time a strong reversal of the ideas of ocean's agency. The imperative exists for Sedna to become clean or, as we might extrapolate, the imperative for the sea—her world, and an expanded correlate of her selfhood—as the theater of the hunt, to be ordered ritually once more, freed of the impurity of taboo violations. Yet Sedna, as *inua* of the sea, cannot, in fact, keep herself unaffected by human behavior; she needs the intermittent help of an intermediary between the worlds to be released from its unending cycles. Her pollution—the pollution of the sea—is semiotically rich, inevitably self-reflexive, and multivalent. It is bound up with human moral and ethical codes and their abrogation. Her role as an outcast, the journey she began when she agreed to become the bride of a bird or a dog, gives her a distance from human society that nevertheless cannot protect her from its internal decay. Her purification—the purification of the sea—undertaken at tremendous risk, is similarly a collective moral reflex, played out in her body—in the culturally charged idiom of dirty hair becoming clean. The ordeal serves, even if only temporarily, to rebalance the society that once spurned her and now desperately needs her grace to hunt, to eat, to live.

CHAPTER 6

"O Ocean, I Ask You to Be Merciful"

The Hindu Submarine Mare-Fire

Themes of gender, its relationship to authority, and the religiously imagined sea revolve kaleidoscopically in the disturbing myths we have just considered. Iphigenia and Sedna are women who are derailed from the anonymous social tracks that should have been their respective lots: marriage, household, children, widowhood, old age, death. That uncanny derailing segregates them and lifts them out of the quotidian world of domestic endeavor, interlaced as it is with the observance of taboo, setting them instead into direct, chronic relationship with supernatural beings and forces, violence, and the ocean. Having failed for one reason or another as brides, both are brutally sacrificed on marine altars by their fearful fathers for the sake of expediency and survival. Both are thus married to a different destiny. After their ordeals, neither can return home to normal land-based lives but instead remain permanent sea dwellers—one at the seaside; one below it. Both women are transformed into ritual experts themselves, with notable religious powers. Exercising those powers in an oceanic environment, both become arbiters of ancient codes of purity and pollution, which are, as we have seen, intimately bound up with traditional ethical systems. The idea of a "breach of taboo" is interchangeable, in Sedna's cycle, with the literal pollution of the body of Sea Woman herself. In ancient Greek religion, concepts of evil, guilt, and crime are symbolic equivalents and are used in literature and inscriptions in apposition with far more concrete terms of pollution such as "stain," "filth," "miasma," "plague," or "malady." Displaced women—Iphigenia and Sedna, priestess and goddess—each involuntarily caught in strange worlds by forces beyond their control—nevertheless eventually determine how these powerful complexes of disease and transgression are to be interpreted and deployed in the worlds they were forced to abandon: the social and religious worlds of ordinary mortals. The patriarchal violence wrought upon Iphigenia and Sedna changes them forever and arguably even infects them, since they themselves are then initiated into violent ways. Yet this violence also, paradoxically, empowers them, transforming them into moral arbiters, keepers of righteousness and ritual law. In each case, the sea not only provides a setting for the female control of pollution, purity, and morality, but also informs the ritual expression of these entities.

An electrifying variant of these themes is the subject of this chapter. It is the medieval Hindu account of an inferno unleashed by one god in response to the torment of another, threatening the entire cosmos, when the male ocean voluntarily swallows the female hellfire of doom, until she is released at the end of the world. This is the Indian mytheme of the burning submarine mare (*vaḍavā*), a creature whom only Ocean, Samudra, can contain and render harmless. According to Wendy Doniger (O'Flaherty), who has studied this complex in depth,

> Fire and water are given varying assignments of gender; but the Vedāntic concept of (male) water as food for female (fire) causes the scales to tip in favor of the distribution in which the woman is the devourer, man the devoured: the fire is a mare. Since rivers are feminine and the ocean masculine, the mare remains a mare when "married" to the ocean.[1]

Although the theme appears in a number of literary contexts from different periods, its most cogent and prominent versions are found in the Purāṇas, traditionally believed to have been uttered by Brahmā, consisting of highly variegated narratives of oral dialogues interpreted by sage-like commentators. The main wellspring of what today might be called, depending on one's allegiances, either Hindu "myth" or "sacred history," the Purāṇas, like the Homeric epics or the Norse prose and poetic Eddas, relate stories that may predate their final literary form by centuries or even millennia, as well as infinite redactions, recombinations, and ideologically motivated representations of those stories. The Purāṇas, however, unlike the Greek epics or the Eddas, dwell almost exclusively on the doings of the gods. Certainly, as we will see, the paradoxical idea of fire contained by water is as old as the Vedas, if not older, and forms the much larger semantic field of which this vivid marine tale is only one expression.

The voluminous Śaivite text *Śiva Purāṇa*, the composition of whose individual Khaṇḍas apparently spans the period between 750 and 1350 C.E.,[2] tells in its lengthy Rudrasaṃhitā of the lethal fire released from Śiva's third eye when he was deluded and aroused by Kāma (Desire). (Virtually identical accounts are given in *Mahābhāgavata Purāṇa* 22–23 and *Kālikā Purāṇa* 44. 124–136). Once loose, Śiva's fire could not return to him, and despite Brahmā's effort to shield Kāma, burned the divine attacker to ashes. The fire began to scorch the triple worlds and to threaten the entire universe. The gods sought refuge with Brahmā.

The Story of the Submarine Fire-Mare

Śiva Purāṇa 2.3.20, 1–23[3]

Nārada said:
1. Brahmā, please tell me, "Where did the flame of fire emerging from the eye of Śiva go?" Please tell me also the further story of the moon-crested lord.

Brahmā said:

2. When the fire from the third eye of Śiva reduced Kāma to ashes it began to blaze all round without burning anything.

3. A great hue and cry was raised in the three worlds consisting of the mobile and immobile creatures. Immediately the gods and sages sought refuge in me.

4. All of them in their agitation bowed to and eulogized me with their palms joined in reverence the heads bent down. They intimated to me their grief.

5. On hearing that I pondered over the reason for the same, and remembering Śiva humbly I went there in order to protect the three worlds.

6. That fire, out to burn everything, very brilliant, with its shooting flames, was thwarted by me as I had the capacity by Śiva's grace.

The gods panicked, and in haste, the great god Brahmā transformed the fire into a female horse. "For the benefit of the worlds," as the text has it, he brought this fiery mare to the edge of the sea.

7. O Sage, then I made that fire of fury, out to burn the three worlds, tender in its blaze and mare-like in its shape.

8. Taking that fire mare-like in form, at the will of Śiva, I, the lord of the worlds, went to the sea shore, for the benefit of the worlds.

9. O sage, on seeing me arrived there, the sea took a human form and approached me with palms joined in reverence.

10. Bowing to and duly eulogizing me, the grandfather of all the worlds, the ocean said lovingly,

The ocean said:

11.–12. "O Brahmā, the lord of all, why have you come here? Please command me with pleasure knowing me to be your servant." On hearing the words of the ocean I remembered Śiva. I spoke with love in order to benefit the world.

Brahmā said:

13. O dear, intelligent one, causing the welfare of all the worlds, O ocean, induced by Śiva's will, I shall explain to you.

14. This is the fire of fury of Lord Śiva, the great lord. It is in the form of a mare now. After burning Kāma it was about to burn everything.

15. At the will of Śiva I was requested by the gods who were harassed by it, and so I went there and suppressed the fire.

16. I gave it the form of a mare. I have brought it here. O ocean, I ask you to be merciful.

Highly apparent is the reverence of the divine sea to its creator, but also what the Hebrew Bible might call the *ḥesed*, or loving-kindness, of Brahmā toward the self-sacrificing Ocean: "O ocean, I ask you to be merciful."

17. This fury of lord Śiva, now in the form of a mare, you will bear until the final dissolution of all living beings.

18. O lord of rivers, when I shall come and stay here, you shall release it. This is Śiva's wonderful fire of fury.

19. Its perpetual diet shall consist of your waters. This shall be preserved by you with effort lest it should go down.

20. Thus requested by me, the ocean agreed. None else could have grasped Śiva's fire of fury thus.

21. That fire in the form of a mare entered the ocean and began to consume the currents of water. It blazed with all its shooting flames.

22. O sage, then, delighted in mind I returned to my abode. The ocean of divine form bowed to me and vanished.

23. O great sage, the entire universe, freed from the fear of that fire became normal. The gods and sages became happy.

Ocean has done his task, accepted his burden. "The ocean of divine form bowed . . . and vanished." *Vaḍavālana*, the doomsday fire in

mare form (elsewhere called "Jvāla Mukhi," or "Mouth of Fire"),
enters the salt waters and is at once checked, surrounded, and
delimited by an endless aquatic sphere. The result is "normalcy," the
possibility of a return to the status quo: "The entire universe, freed
from the fear of that fire became normal. . . . The gods and sages
became happy."

In this text, the mare seems dwarfed by the vastness of her
marine prison as she grazes on the salty waters surrounding her. The
fire eats; the water is eaten yet is never diminished because the sea
is inexhaustible. We imagine her ceaselessly wandering along the
sea's darkest depths, luminescent, with a soft blue light around her
muzzle, "cool" or "pleasant" (*saumya*) like the moon. This ambro-
sial fire in one sense reiterates Śiva's own halo of flames, the *prabha-
maṇḍala*, the encompassing circle of fiery light that surrounds him in
his role and image as Naṭarāja, "King of Dancers," and terminates in
the prostrate body of the dwarf of illusion, Apasmāra Puruṣa. In another sense, however, the fire of the mare represents only the potential
of divine energy's release, not its theophanic expression. Deceptively
silent, flickering at the bottom of the vast sea, the mare-fire can ruin
the whole world if Ocean ever lets her go. The story makes clear that
one day he will. Concealed is her tremendous *mana*, the inferno of
Śiva's wrath, enough to scorch all the gods and the universe in its
entirety.

Just as the *Śiva Purāṣa* henotheistically represents one god, Śiva,
as comprising and thus subsuming in his *saguṇa* form three other
great gods (Brahmā, Viṣṇu, Rudra)—at least two of whom, in
other texts with different allegiances, may subsume him—exclusive
claims of salvific efficacy are made for one divine entity, Ocean.
In ancient Greece the sea was believed to be more powerful than
any other body of water for its cathartic potential. So in this myth
no river, not even Gaṅgā, is great enough or pure enough to over-
power and neutralize the scalding power of the Śiva-fire, the mare.
It is only Ocean, "lord of rivers," who can permanently contain the
scalding fire, at least until the end of time: "None else could have

grasped Śiva's fire of fury thus." Ocean compassionately stoppers the ravenous one who could decimate the cosmos, absorbing her into his own body. In so doing, he renders life on land possible once more. As in the case of the plague-ravaged Theban chorus members in *Oedipus the King* who pray for the scourge to be driven into the sea, the collective desire is both simple and utterly familiar in the contemporary world—*for life to continue as it was before the communal threat arose.* "The entire universe, freed from the fear of that fire became normal. The gods and sages became happy" (*Śiva Purāṇa* 2.3.20.23).

Themes of Fire and Water

Like birthplace and graveyard, paradise and hell, creation and destruction, or eroticism and asceticism, fire and water are unyielding oppositional pairs, the kind to which mythic thought and the theological imagination are magnetically drawn, expressing their strange relationship in sacred history, ritual, and iconography. The more polarized the difference between important conditions or entities, the more myth insists on marrying them, revealing one as the shocking but inevitable culmination of the other—the hidden heart of its extremity. In this trajectory, rather than annihilating the other, one extreme will, in time, *regenerate it* instead. Jaan Puhvel remarks on the counterintuitive union of fire and water in Indo-European religious thought, especially the role of their inseparable duality in ideas about the beginning and ending of the world:

> Fire and water are archetypally antithetical in the physical world and in human perception alike. In the former their incompatibility is relentless, but in the mind of mythopoeic man it has created its own dialectic of conflict resolution that is reflected in ancient tradition. Fire and ice/water were both present in Norse cosmogony and eschatology alike. "Fire in water" is a theme that recurs in

Indo-Iranian, Irish, and Roman lore, in a complex mythologem of clear Indo-European relevance.[4]

As Wendy Doniger emphasizes in a number of her works dealing with this theme, fire and water have always been ritual purifiers in the Indian religious world. They are held in perfect, balanced tension in the myth of the submarine mare.[5] This marriage of opposites is anticipated by the Vedic hymns about the identification of Agni, "the one sea with many births" (*Ṛg Veda* 10.5) with "the Child of the Waters" (*Apām Napāt, Ṛg Veda* 2.35), born in the sea, who, golden, enters the sacrificial fire and becomes one with him.[6] In *Ṛg Veda* 10.51, by hiding in the waters, the place of his birth, Agni (Fire) escapes his onerous role as "oblation giver," vehicle of the sacrifice, a horse harnessed to his task. But the gods track him down there with his "many-coloured light which shines beyond the distance of ten days' journey" (3) and persuade him to return by promising him a long life span free from old age.[7] Agni, divinized Fire whose home is in lightning and the body, as well as in the many forms of *yajña*, is a seminal golden force without whose work, "carrying the offerings," the ritually driven cosmos would cease to unfold: "Man, who loves the gods, wishes to sacrifice" (*Ṛg Veda* 10.51.5). The waters do not extinguish Agni either at birth or in his flight from his sacrificial vocation; rather, this dialectical relationship between fire and water appears to be a very early one in the religious imagination of ancient India.

The ostensibly later story of the mare-fire, part of the biography of a non-Vedic god, Śiva, may be related to the Vedic dialectic between these two elements. However, Vaḍavālana belongs to a different branch of the family tree of symbolic thought about fire-in-water. In no way does she participate in religious culture. Unlike Agni, she is not sacrificial; she is only hellacious. Unlike Agni, she is not a god in her own right but is rather the hypostasis of only one aspect of a mighty god, divorced from his total divine nature that encompasses so many seemingly opposite aspects. Just as each of the ancient Greek gods embodies contradictory realms of power—Artemis is goddess of both wild, sylvan virginity and the domestic, sexually mature sphere of childbirth; Apollo brings both plague and healing, since both are idiosyncratically his—so Hindu gods also are dialectical in nature. Śiva, the erotic ascetic, creator and destroyer, oscillates in myth according to the situation yet remains devotionally and philosophically in perpetual integrity, owning every aspect of himself as an atom binds its whirling neutrons in their orbits. The myth severs the destructive capability of the god's third eye from the paradoxical yet integral godhead and sends it to live out its own destiny. While supernatural, then, the fire is also unnatural. Mutating from its Indo-European ancestor—the tale of the deity who, in Puhvel's words, "hoards a fiery and effulgent power immersed in a body of water"[8]—Vaḍavālana, although divinely sequestered in the sea, is neither valuable nor precious.

Cut off from its source, the fire of Śivā-hastily-turned-mare might easily devour all she encounters, even the gods themselves. Because she is fire, all is food for her, and thus she must be given the whole sea to eat, the only possible "pacifier." In some versions, clearly related to the one we are considering, the mare originates from different divine or divinely made beings, such as the *Brahmā Purāṇa* story of the mare-shaped wrath of the sage Pippalāda, who, when he was able to see the third eye of Śiva, also "gained from Śiva the power to kill the gods." Taunted by the fig trees that his mother was a mare, Pippalāda's anger issues from his eye in the form of a mare (110. 85–210):

Then the mare set out full of fire to burn the universe, terrifying the gods, who sought refuge with Pippalāda; but Pippalāda could not restrain the mare. When she came to the confluence of the Ganges, the gods begged her to begin with the waters of the ocean and then to devour everything. The fire said, "How can I reach the ocean?

Let a virtuous maiden place me in a golden pot and lead me there."
The gods asked the maiden Sarasvatī to do this, and she asked them
to join her with four other rivers, the Yamunā, Ganges, Narmadā,
and Tapatī. The five rivers put the fire in a golden pot and brought
it to the ocean; they threw it into the ocean, and it began to drink
the waters little by little.[9]

As watery conduits, the rivers can bear the mare-fire away from
society as far the ocean, but they cannot offer a permanent solution.
Again, that is the special task of the sea. The gods trick the mare-
fire into holding off her voracious grazing until she can start with
marine waters. They know that she will never reach the end of eating
those waters, and so they will be safe. In other versions, which omit
the role of the rivers, the mare is given in a golden pot from a hermit-
age to Sarasvatī by the gods, who charge her with its sequestering.
"You must take this mare-fire and throw it into the ocean of salt so
that the gods will be relieved of their fear; otherwise the mare-fire
will burn everything with its fiery energy."[10]

The tale also belongs to a larger family of later Hindu myths of
the containment of burning, destructive substances by waters, such
as those pertaining to the transformation of the fire of Śiva into the
demon Jalandhara at the confluence of the Ganges and the sea, or the
absorption of Śiva's burning seed by the River Ganga, or the similar
story of the birth of six-faced Kārttikeya (Skanda).

> Once Agni drank the spilled semen of the Pināka-bearing god (Śiva),
> O brahmin, and was overcome by it. Deprived of his strength, that
> infinitely shining god then went for help to the gods, who sent him
> at once to Brahmaloka. On his way, Fire met the goddess Kuṭilā.
> When he saw her, he said, "O Kuṭilā, this seed is most difficult
> to bear! Since it was spilled by Maheśvara, it would burn up the
> three worlds if uncontained. Therefore you take it! The offspring
> you bear from it would be auspicious."[11]

The *Vāmana Purāṇa* recounts how the mighty river Kuṭila, eager to bear a child, did allow Agni, burnished gold by his long pregnancy, to hurl the burning seed of Śiva into her waters. The river's own five-thousand-year pregnancy drove her in distress to Brahmā, and having pity on her, the god directed her to an enormous reed-thicket on Sunrise Mountain where she might release the embryo. After ten thousand autumns the child Skanda was born, attracting the jealous attention of the six shining Kṛttikās, the sister stars the Pleiades, part of Ursa Major, for each of whom the baby produced a separate face. Here the pregnancy of Kuṭila, the fresh river, parallels in certain ways the role of Samudra, the salt ocean, in the Śivaite myth. Samudra will also contain the mare-fire, an odd embryo of sorts, for a bounded, albeit much longer, period of time, until her "birth": the death of the world.

It is especially ironic, or perhaps it is just one more of the paradoxical relationships treated earlier, that the fire that threatens the world—the fire that must be contained in the sea—originates with Śiva. In many accounts, among them the *Rāmāyaṇa*, the *Mahābhārata*, and the *Viṣṇu Purāṇa*, it is he who autosacrificially (perhaps shamanistically?) once drank the sea's deadly poison, thereby saving the world.[12] The poison was produced when the gods churned the primordial Samudra of milk in order to produce the elixir of immortality. As Jeanine Miller comments, "Both the poison and its antidote are contained in the ocean or substance of divinity which in the past produced everything, both good and evil, both life and death. This explains why endeavor to obtain life is threatened with being turned into a receiving of death."[13]

The specific theme of the submarine mare, found in the *Śiva Purāṇa* and in the early poetry collected in the *Subhāṣitaratnakoṣa*, differs both from such stories of fire held in water in its insistence on the uniqueness of Ocean's role and in its clear apocalypticism. The myth appears in various forms in the exquisite collection of Sanskrit poetry, dating as far back as the fifth century and earlier, discovered by the eleventh-century Buddhist monk Vidyākara in the library of his monastery at Jagaddala in east Bengal and anthologized.[14]

> How marvelous the underwater fire!
> How marvelous the blessed sea!
> The mind grows dizzy thinking of their greatness.
> The first keeps drinking greedily its dwelling
> and yet its thirst by water is not quenched;
> the other is so great it never suffers
> the slightest loss of water in extent.
>
> *Subhāṣitaratnakoṣa* 1198

The miracle of the burning submarine fire or mare-fire is that she does not diminish the infinite waters of the sea. Furthermore, elsewhere the hearer is reminded that through its containment and check of the *vāḍava*, the sea, as it is portrayed in the *Śiva Purāṇa*, is saving the world out of universal compassion.

This verse by the Sanskrit poet Keśaṭa is grouped with others that deal with what Ingalls calls "greatness," with the ocean as a primordial image of and metaphor for awesome magnitude. The sea is the birthplace of Śri, the goddess of good fortune and prosperity (1196, 1997), and there she dwells with Viṣṇu (1199), which the sea also bore, along with the cosmic serpent Śeṣa (1195, 1209).[15] The great ascetic Agastya was nevertheless able to drink the very ocean, "that acme of all things that men call marvelous," in Vācaspati's poem (1201), a theme repeated many times in Sanskrit and Tamil, most centrally in the *Mahābhārata*.[16]

More subtle in these poems than Agastya's "maw wherein the roaring ocean rolled / together with its fish and crocodiles and sharks" (1200), but far more ominous, is the sea's apocalyptic threat. The incomprehensibly vast body of water, biotic, generative of the gods (as in the ancient Greek view) also contains the end of life on earth, "an island" that it dwarfs in comparison. The sea, as it rolls, thus bears both the beginning and ending of all things:

> How shall we tell the fortune of the sea,
> The very birthplace of the goddess Fortune?
> Or how express its magnitude,
> Of whom the earth forms but an island?
> What is its generosity, whose suppliants,
> The clouds, support the earth?
> How tell its power, when by its wrath
> We know the world will perish?
>
> *Subhāṣitaratnakoṣa* 1196

The Fire at the End of the World

In the words of the *Śiva Purāṇa*, the submarine mare contains "the final dissolution of all living beings." She is not merely a destructive power that is narrowly contained; she will also unleash the end of the world when she finally emerges from the sea into which she was sent by Brahmā. The marriage of fire and water will be undone, as will their *coincidentia oppositorum*. What, then, are we to make of this mythic idea and, in particular, the role played by Ocean in its various forms?

Drawing from a wealth of Indo-European mythological comparanda, Doniger describes the androgynous and hence unstable mare, with both female maternal and male divine/erotic male qualities, and thus carrying a dangerous, even demonic charge in the texts through which the horse blazes. In contrast to the highly fertile Indo-European mare to which she is related, the submarine mare ought to be read as "a symbol of angry, thwarted [female] sexuality, of power blocked by authority."[17] Why? As Doniger eloquently points out, the fire of Śiva's third eye is an antierotic fire, blazing forth to counter the tormenting, erotic fire. As it is referred to in the

Matsya Purāṇa (154. 250–255), "the 'fire of Kāma' is a pun, denoting the fire used by Kāma and against him as well, a flame composed of two sparks. . . . This is the fire that Śiva places in the ocean in the form of a mare, the fire of thwarted passion."[18] She finds persuasive support for this deliberate conflation—or conflagration—in the poetic combination of Śiva's angry fire, the mare-fire, and the fire of Kāma "in a verse addressed to Kāma *(Abhijñānaśākuntala* 3.2. alt.): 'Surely the fire of Śiva's anger still burns in you today / like the fire of the mare in the ocean; for how else, Kāma, could you be so hot / as to reduce people to ashes?' "[19] Thus, Doniger asserts that "the mare is a symbol of yogic power thwarted and rebounding against itself. . . . The combination of anger and lust is at the heart of the myth of Śiva and the mare."[20]

Doniger's exegesis of the myth of vaḍavālana requires that we accept the gender transformation of Śiva's fiery male eroticism/anti-eroticism into a female mare, which indeed the myth itself stipulates. But are we then to understand the mare's erotic charge in terms no longer "mythical" but now socially constructed as fully, aggressively sexualized female? Is the submarine mare in her apocalyptic potential no longer a divine horse, no longer a severed aspect of the godhead but, rather, a medieval poetic image of human woman, writ as socially dangerous and cast in the tropes of mythohistory? Can *this* particular charge destroy the triple worlds and all the gods? Doniger believes that it can: "In the Hindu view, a woman's suppressed or repulsed eroticism is as volatile and explosive as nitroglycerine. We say that Hell hath no fury to match this, and the Hindus say that this *is* the Fury that breaks forth out of Hell at doomsday."[21]

Pace the recent critiques by Rajiv Malhotra[22] and others of Doniger's work, readily attacked in Hindu neofundamentalist agendas, it is not the case that highlighting the transpersonal sexual resonances in myth is equivalent to "psychoanalyzing" or dishonoring them. The aims of classical psychoanalysis, namely, the methodical retrieval of unacceptable unconscious contents, often repressed in personal history, that impede the individual's ability to enjoy full psychological freedom, are in no way comparable. One of the tasks of the historian of religion is to reveal shared or related aspects of traditional mythemes, ideologies, or ritual practices and somehow to account for those interlocking surfaces, either historically or in other ways. Differences in mythemes are unsurprising. Similarities, surprising and ubiquitous, are what cry out for explanatory thought. Such thought Doniger has repeatedly brought to bear on myriad cases of mythologies from around the world, not just Hindu. These shared aspects may or may not be desirably discernible or deemed worthy of theological commentary by scholars within the tradition. Myth-complexes alone are inexhaustibly multivalent, however, and to lift up one way in which they can be understood, one way in which they refract cosmic or existential concerns, does not mean that others, which can be equally historically compelling, must be shut down. Otherwise we are left only with internal apologetic, which has its value but which by itself cannot fully illumine, invested as it is in particular self-representations of religious tradition.

Thus Doniger's analysis of the significance of the mare as an archetype of "the suppressed woman," even in its apocalyptic role, can be defended for the Śaivite forms of the myth, since following such a thread does not distort. In the account of the *Śiva Purāṇa*, the mare-fire indeed does seem to represent the angry blaze of the frustrated god-yogin, passionate (when not frustrated) in both his generation of *tapas* and in making love to Pārvatī. As we have seen, however, Doniger must account for the complete conversion of male divine energy into human female energy and must therefore stress a gendered, sociosexual reading of the myth.

There are also Vaiṣṇavite stories of the mare's role at the end of the world, as in the *Matsya Purāṇa*, which Doniger herself cites:

Viṣṇu becomes the sun and dries up the ocean with his scorching rays. . . . He goes down to Hell and drinks the water there, and then he sucks out the urine, blood, and other moisture in the bodies of living creatures. . . . He is the mare's head in the ocean of milk, the whirlpool fire that drinks the oblation made of water.

Matsya Purāṇa 166.1–4; 167.58–59

In a text like this, is it that "Viṣṇu/Kalki plays the role of the mare/woman"? Or is the myth of the mare-fire in the sea perhaps more protean than the fixed, particular construction that Doniger O'Flaherty wants to give it? In light of other non-Śaivite apocalyptic accounts, might the story of the mare-fire not be read as an expression of the cataclysmic uncoupling of the larger dialectic of fire and water? In the Purāṇic vision, fire is always in one way or another eschatologically brought to bear on the cosmos (e.g., released from the sea by Brahmā, borne from the seven suns permeated by Viṣṇu, and so forth). In the *Viṣṇu Purāṇa*, for example, the "monstrous fire" brings the last Kali Yuga of a thousand cycles to an end, causing the three worlds to "blaze like a frying pan"[23]—realizing, in other words, the same fear that the gods express while the fire from Śiva's third eye rages until it is turned into a mare and given to Ocean to swallow and to be swallowed by. The inferno blazes until Rudra produces elephantine, jewel-colored clouds that "completely extinguish this dreadful fire" and keep pouring until the world is covered once more in a primordial sea, so that "everything movable and immovable in the world has perished in the watery darkness." Viṣṇu reposes once more on the great serpent Śeṣa, "in the single ocean . . . resting in meditative sleep, in the divine form of his own illusive power." This oceanic sleep continues through the night, a night as long as the "day of Brahmā"—the cycle of ages just terminated—until the dawn when Viṣṇu, "unborn," shall awaken to take the form of Brahmā to recreate the world, a world in which fire can once again be born in water or be confined in it. In other words, the "unleashing" of fire, of which the mare-fire is only one form, and the water, of which the sea flood is an intensified expression, is the template for the end of time and also the prelude to its recreation.

As I wrote these words at the end of the year 2004, countless funeral pyres burned along the beaches of southern India: bright, hungry fires against the backdrop of the quiet, occulted sea, that then had yet to give up so many of its dead to those who searched for them. "How shall we tell the fortune of the sea / . . . How tell its power, when by its wrath / We know the world will perish?"

CHAPTER 7

"Here End the Works of the Sea, the Works of Love"

Εδώ τελειώνουν τά ἔργα τῆς θάλασσας, τά ἔργα τῆς ἀγάπης
(Here end the works of the sea, the works of love.)

George Seferis, *Mythistórima*

Apparently "freed from that fire" of our own toxic by-products—that of which we would be rid forever—like the gods and sages we too are "happy" that the ocean has relieved us of them. Only recently has it come to our attention that, as the Purāṇas tell us, the sea might not always embrace that which could destroy the world. What went into its depths has the potential to emerge once more, when the time comes.

"O lord of rivers, when I shall come and stay here, you shall release it." Brahmā's words to Ocean forecast a grim ending to the episode of the loosing and containment of the fire of Śiva. The "relief" that the gods and sages feel at being rid of this scourge is overcast: selfless Ocean will not be able to accommodate the underwater mare forever. Brahmā says, "Now you must bear it until the final deluge, at which time I will come to her and lead it away from you." As we have seen, the mare's destructive potential will remain untapped and her mare-form unchanged until Brahmā himself decrees the end in the Fourth Age, the Kali Yuga, and in the Purāṇic vision, Viṣṇu as Rudra will allow the sun to scorch the entire earth until it is "bare as a turtle's back."[1] The torrential rain clouds sent by Viṣṇu will quench the fire; only the ocean, absorbing all created beings into itself, will be left. Upon that primordial sea on the back of Śeṣa, Viṣṇu will sleep peacefully until he again arises as Brahmā to recreate the cosmos. The sea will be transformed from graveyard to birthplace once more until it reverts to graveyard.

The solution to the acute problem of the apocalyptic fire—its theriomorphic change, its submersion in Ocean—although apparently permanent, will instead be temporary and therefore unstable. The Purāṇic account of the salvation of the world also contains the narrative of its ultimate destruction. Divine, ordering power expresses a limit to the sea's ability to absorb menace. Thus that limit, like the circumscription of historical time itself, is divinely ordained. At the end of this age, by the will of Brahmā, what Ocean bore and neutralized for the sake of the world will be unleashed, and what had been done will be undone.[2]

In the preceding three short studies, we saw culturally and religiously determined ideas of the ocean as the place that can always cleanse or absorb certain evil threatening humankind, whether ritually produced by humans themselves (as in the ancient Greek ritual system and the myths of Sedna) or by sacred processes run amok (the fire of Śiva's third eye). However, what is also clear is that a number of disquieting myths around the world also contain the idea of the *reversal* of this process. This element of the mythical imagination is nightmarish: the sea releasing, surrendering, or regurgitating all that it has absorbed to allow the human *oikoumenē* to function free of its own toxins. This reversal is frequently linked to the end of time, as is the case of the Śiva Purāṇa. These narratives warn that there may also be an unimaginable time when at last what went into the sea, apparently discarded forever, might reemerge from it at the final reckoning. The fiction of irretrievability— that things "disappear" into the ocean and never reappear—is itself dissolved in such accounts, inspiring the same peculiar dread produced by the motif of the dead and buried rising out of their graves.

In the vision of *Moby Dick*, all the ocean is a great crypt, and an unquiet one at that:

> When gliding by the Bashee isles we emerged at last upon the great South Sea; were it not for other things, I could have greeted my dear Pacific with uncounted thanks, for now the long supplication of my youth was answered; that serene ocean rolled eastwards from me a thousand leagues of blue.
>
> There is, one know not what sweet mystery about this sea, whose gently awful stirrings seem to speak of some hidden soul beneath; like those fabled undulations of the Ephesian sod over the buried Evangelist St. John. And meet it is, that over these sea-pastures, wide-rolling prairies, and Potters' Fields of all four continents, the waves should rise and fall, and ebb and flow unceasingly; for here, shades of mixed shades and shadows, drowned dreams, somnambu-lisms, reveries; all that we call lives and souls, lie dreaming, dream-ing, still; tossing like slumberers in their beds; the ever-rolling waves but made so by their restlessness.[3]

Here are echoes of Faulker's yearning for the sea and the rush of relief and joy he feels when he meets it after a long separation, stretching to the horizon; but here too is the vast unease occasioned by a sailor's awareness of the lives lost beneath the ocean's chaotic surface. Those lost lives are far from at peace; Ishmael imagines them not in deep but rather in restless slumber, drenched still in dreams, not properly buried, suspended in some kind of watery *entrepôt*. The sea becomes a vast, universal potters' field, a wild graveyard for those without a true "place" in which to be interred. Its very waves, its chronic motion, are, in Melville, actually caused by the tossing and turning of the drowned—un-still, unreleased, un-free. The sea does not make the dead roll; they make *it* roll instead.

"Ding-Dong, Bell"

The idea of the great ocean's surrendering its dead—in a sense, rejecting them—encodes a kind of mythic inversion signaling the reversal of creation and of the normal order of things. The sea receives all perishable things and changes them completely, rendering them unrecognizable: Shakespeare's "sea-change / into something rich and strange." This is most dramatic in the case of the dead.

The dumping of corpses into the sea in an effort to erase not only the manner of death but even any identifying physical characteristics is an age-old ploy, a favorite of criminals played out to this day not only in literature but also in the headlines. A death at sea always threatens anonymity and the erasure of one's place in collective memory, as the remaining lines of the Lord Byron stanza from "Childe Harold's Pilgrimage" cited in chapter 1 testify:

. . . —upon the watery plain
The wrecks are all thy deed, nor doth remain
A shadow of man's ravage, save his own,
When, for a moment, like a drop of rain,
He winks into thy depths with bubbling groan,
Without a grave, unknell'd, uncoffin'd, and unknown.

This linkage of sea and anonymity at death is crucial to Macrobius's Carthaginian martyrologies, in which the Romans are frustrated that the execution of Christians produces only an increasing supply of relics to venerate, thus strengthening the cult. They decide to load Christian bodies together with those of executed criminals into boats, row them out in the Mediterranean, and dump them all into the sea.[4] The idea is to annihilate the martyrs' bodies along with their identities using the powers of the hungry sea, and thus to halt the generation of new relics. As is typical of early Christian saints' lives from every geographical area, the evil plot is foiled by the miraculous intervention of animals who protect holy persons living and dead.[5]

In this case they are sea creatures. As Maureen Tilley describes the "plot" of this account,

> Even if the tide eventually washes the bodies back to shore, the remains of the martyrs will be so disfigured by several days in the sea and so intermingled with the bodies of common criminals that the Christians will be unable to tell the difference. They will refrain from venerating any of the bodies for fear of honoring murderers and robbers. . . . Immediately dolphins come to the aid of the Christians and bring back only the bodies of their revered saints before the sea can take its toll.[6]

The sea's tendency first to disfigure and then, if left to its own devices, to obliterate the identities of the corpses has been observed (and exploited, as the Macrobius story shows) throughout time. Going one step further still, the sea can transform a corpse into something sealike and magical, for after the process of decomposition is over, only the filigree of bones are left. This is the source of the rhetorical strength of Ariel's famous (and false) dirge for Ferdinand's father in *The Tempest* (act 1, scene 2):

Full fathom five thy father lies;
Of his bones are coral made;
Those are pearls that were his eyes;
Nothing of him that does fade,
But doth suffer a sea-change
Into something rich and strange.
Sea-nymphs hourly ring his knell:
Ding-dong,
Hark! Now I hear them—Ding-dong, bell.

But what if the ocean's dead will not stay put? A classic of early Indian Mādhyamika philosophy, Candakīrti's *Madhyamakāvatāra* (*The Entry into the Middle Way*), compares the pure nature of the bodhisattva at the "second stage in the generation of the thought of

awakening" to the ultrapure nature of the sea, which will not brook
a corpse: "Just as in the case of the ocean with respect to a corpse
or as it is with prosperity in the face of misfortune—so a mighty
one (*mahātman*) governed by the force of morality is unwilling to
live with any transgression."[7] In this strain of thought, with parallels
in early Pāli monastic texts (e.g., "As the ocean rejects a corpse, so
the monkhood rejects evil-doers"), the sea's ritually overdetermined
purity will not allow it to retain any entity as impure as a corpse.
This physical, observable marine quality is seamlessly rendered as
metaphysical. The ocean becomes an emblem in the natural world
of the evolution of the aspirant on the path to final awakening and
liberation. Corpses may enter the sea, but they cannot stay there for
long, because the sea, like the mind of the immaculate (*vimalā*), will
self-purify, ridding itself in a reflexive act of agency of anything that
is polluting and thus foreign to its nature.

Like the story of the final release of the submerged, burning
vaḍavā from Ocean at the end of time, the book of Revelation,
the Christian apocalyptic text of the Imperial period that concludes
the New Testament, imagines such a reversal. The sea is the source
of the great beast in Revelation 13:1, who rises out of it to join the
dragon taking his stand "on the sand of the seashore" (Rv 12:18) in
the eschatological combat. The sea beast, "having ten horns and
seven heads; and on its horns were ten diadems, and on its heads
were blasphemous names," combines the characteristics of each of
the four beasts in the vision of Daniel 7.[8] In Revelation 16:3, the
sea becomes the receptacle for one of the seven bowls of the wrath
of God, and is transformed thereby into a vast watery killing field:
"The second angel poured his bowl into the sea, and it became like
the blood of a corpse, and every living thing in the sea died."

Yet after Christ's defeat of Satan the beast and the nations "as
numerous as the sands of the sea" (Rv 20:7) at the end of the
millennium-long interregnum, an even stranger sign transpires.
This event signifies a reversal of all that has gone before and the in-
auguration of the eternal heavenly kingdom by physical resurrection.
"The sea gave up the dead that were in it, Death and Hades gave up

the dead that were in them, and all were judged according to what they had done" (Rv 20:13). The passage is clearly based on Jewish apocalyptic ideas set forth in 1 Enoch (the Ethopic Apocalypse of Enoch), a text dated between the second century B.C.E. and the first century C.E.[9] The retrieval of the dead for the Last Judgment is not confined to earthly graves or to the infernal realm but will include—in fact, will begin—with a ghastly display of the ocean's hidden treasuries. "Death and Hades" are shown to be but temporary holding places for the dead, but so, surprisingly, is the sea. As Wilfrid Harrington remarks, "It was widely believed that those lost at sea had no access to Sheol" (Hades).[10] Revelation addresses that belief directly and offers remedy. The book foresees that all the righteous dead, even those from the vast crypt of the sea, will return on the day of Resurrection. Moreover, like Ocean in the medieval Hindu story, the sea of Revelation will *surrender* what it hides, an action that will trip a much larger dramatic sequence. Like the Purāṇic account, this marine release of what been submarine for ages will be both sign and start of the end time. The Last Judgment will catalyze the passing away of the first heaven and the first earth, the arrival of "a new heaven and a new earth" in chapter 21, and the descent of the heavenly Jerusalem.

It will also be the end of the sea itself. The next time the sea is mentioned in Revelation after its surrender of its dead, it is as a lost element of the old world. The annihilation of the sea is the first observation about the divinely won eschaton: "And I saw a new heaven and a new earth; for the first heaven and the first earth had passed away; and the sea was no more" [καὶ ἡ θάλασσα οὐκ ἔστιν ἔτι] (Rv 21:1). The final triumph over the sea at the end of time harks back to the containment of the primordial sea by Yahweh in Genesis, which was, as we have seen, *not* final. Adele Yarbro Collins writes,

This statement is parallel to the eternal confinement and punishment of Satan, the beast and the false prophet. It is also analogous to the remark that "Death and Hades were thrown into the lake of fire" (20:14). The *sea*, like the dragon and the beast, symbolizes chaos. So the elimination of the *sea* symbolizes the complete triumph of creation over chaos, just as the elimination of death implies the complete victory of life over death.[11]

All the cosmic checks and balances put in place at the beginning of time are thus upended. Among them is the sea, the "not-land" which, once it has given up its treasury of dead at the Resurrection, can be neither redeemed nor transfigured.

The world-destroying fire of the Hindu account and the dead of New Testament apocalyptic are very different in nature; but they are aspectually similar in function. They both ended up in the sea, and this "saved everything." This immersion, this swallowing, allowed terrestrial human existence—the world of the living—to continue normally in an uncontaminated and hence unthreatened state. In both cases, the immersion is not permanent. However, its "undoing"—the reversal in which the sea gives up the horror from which it has sheltered humankind for ages—is not simply dangerous; it is cataclysmic and is associated with the end of everything once saved by the ocean's containment.

It is significant as well that in both scenarios, the sea itself is portrayed as voluntarily releasing the anomalous entity—the mare-fire or the dead. When the time comes, Brahmā tells Ocean, "you shall release it." And in Revelation, "the sea gave up the dead that were in it." Surely, like the similar regurgitation by Death and Hades, like Ocean in the *Śiva Purāṇa*, the sea does so in response to divine, adamantine will. Nevertheless, the sea's own agency, its cooperation with that will, is inferred. It gives up what it has sequestered and does not need to be coerced or plundered. Nothing needs to be extracted from the cooperative sea. Brahmā will not need to go down into Samudra's depths to lead out the submarine mare. In Revelation or elsewhere, there is no ancient Christian tradition of the harrowing of the sea, as there is of Hades. We might ask whether in the religious imagination, the sea ultimately *must* "surface" what was foreign to it.[12] The return of the dead, from the sea and elsewhere, will begin the end of the painful separation of the heavenly and earthly dominions, the end of death itself:

> See, the home of God is with mortals. He will dwell with them; and they will be his peoples, and God himself will be with them; he will wipe away every tear from their eyes. Death will be no more; mourning and crying and pain will be no more, for the first things have passed away. (Rv 21:3–4)

Echoes of this basic mythic reversal can be found in various literary accounts, both explicitly religious and secular. No matter what the narrative vehicle, the theme of things working backward is at the least unnerving and at the worst terrifying. All three of the following literary works are fed by the stream of this idea: the sea releases and thus exposes any moral transgression that was deliberately cached in its depths. The sea demands a reckoning.

Gershon's Monster

One of the earliest Hasidic wonder tales is spun from the Jewish custom of casting one's sins into the sea (*tashlikh*) at the New Year. Discussed in chapter 3, this observance takes place on the first afternoon of Rosh Hashanah when people carry bread crumbs, symbolizing the sins of the past year, to the sea or to lakes or rivers (if the sea, the best destination for purification, is too far away). There they recite biblical verses of repentance, beginning the process of *t'shuvah*, literally meaning "return." Admission of wrongdoing, remorse, resolution never to repeat the wrongdoing, restitution, and asking the forgiveness of those distressed or harmed are the required steps of *t'shuvah*. Interior change and self-humbling actions of reparation must accompany the ritual of casting sins into the sea. The wonder tale is a critique of external religiosity absent true repentance, and its vehicle is the cathartic ocean that marries ritual and moral cleansing and rejects their divorce.

Retold most recently by Eric Kimmel, whose ancestors came from Constantsa in Poland on the shores of the Black Sea where the story takes place, "Gershon's Monster" is about a baker named Gershon who lives with his wife Fayga.[13] He is an important man in town, although sadly he and Fayga are childless. Gershon is no different from most people in that he makes common, inconsequential mistakes: a lie here, an unkindness there, a broken promise. But he is very different from most others in that he never feels regret for what he has done nor asks anyone's forgiveness. Every Shabbat evening, Gershon sweeps his mistakes and thoughtlessness down into the cellar. "Then, once a year on Rosh Hashanah, he stuffed them into a sack, dragged the enormous bundle down to the sea, and tossed it in. But selfishness and thoughtless deed are never disposed of so easily. There is always a price to pay, as Gershon was about to learn."

In despair over his childlessness, Gershon responds to his wife's news of a *tzaddik* in Kuty and travels by horse and wagon to find him to ask for help. He offends people as he goes and then, when he arrives, barges into the wonder rabbi's house without knocking. The rabbi immediately recognizes Gershon's moral bankruptcy, but for Fayga's sake, begins to pray. Finally he opens his eyes and tells a fidgeting Gershon to be grateful for all he has and not to ask for more. Gershon is furious and demands to know at least why God will not give him a child.

> The tzaddik's eyes searched the depths of Gershon's soul. "Did you think you could live so thoughtlessly forever? The sea cries out because you have polluted her waters! God is angry with you. Accept God's judgement. Your recklessness will bring your children more sorrow than you can imagine." "I will take that risk," Gershon said selfishly.

The *tzaddik* in the story is based on the historical Rabbi Israel ben Eliezer (ca. 1700–1760), the founder of Hasidism also known as the

Ba'al Shem Tov—"Master of the Good Name." The Ba'al Shem Tov was a worker of wonders, battler of demons, and wielder of powers apotropaic, necromantic, and theurgic. The story's *tzaddik* of Kuty gives Gershon an amulet for Fayga to wear around her neck and a prophecy that sounds more like a curse: Gershon's children, a boy and a girl, will be with him for only five years; Gershon will be unable to protect them when they go down to the sea on the morning of their fifth birthday. When the unhappy man begs for a sign, the rabbi warns him of "the day you put two stockings on one foot and storm around the house looking for the missing stocking." He also predicts that once Gershon returns home, he will not in fact remember the sign, nor anything else of the omen.

Fayga wears the charm and a year later gives birth to beautiful twins, a boy and a girl, Joseph and Sarah. They grow up healthy and strong, spending their summer days at the beach. "And Gershon went on behaving as restlessly as ever, sweeping his thoughtless acts into the cellar. And once a year, on Rosh Hashanah, he stuffed them into a sack and dragged them down to the sea."

Five years pass, and Gershon awakes one heavy August morning. In the swimming heat, he pulls his right stocking over his already stockinged left foot. He storms around the house demanding to know who has taken his stocking. When Fayga laughingly points out his doubly clad foot, Gershon remembers the *tzaddik*'s words. Pale and frantic, he races for the seashore, crying out to God in anguish lest he be too late. But he doesn't make it in time. His cries to his children to come away from the water fail to carry, as he is now out of breath. They only wave and resume playing.

When Sarah chases Joseph into the water,

> all at once the sky grew dark, as if a cloud had covered the sun. But it was no cloud. Gershon saw it rising from the sea: an immense black monster covered with scales like iron plates. On each scale was written one of Gershon's misdeeds. "Father! Save us!" the children cried as the monster came toward them.

Gershon runs "as he had never run before" and pushes the children aside. He throws himself before the monster, and "looking up into the creature's glittering eyes, he pleaded for forgiveness."

This moment of reckoning in the tale comes from the sea. Astonishingly, it occurs not between God and man (judge and condemned) but instead between that which has so terrifyingly returned—the monster, hypostasis of years of sins—and the perpetrator, the one who originally and thoughtlessly discarded those sins into the same sea but did not repent of them, thinking never to see them again. Gershon does not ask God for forgiveness, although that is implied, as it is certainly by God's agency that the sins coalesce in this dreadful form. "I know what you are. You are my pride and selfishness coming back to me, just as the *tzaddik* foretold. Please have mercy. Spare my children. Why punish them? Take me instead."

The sacrificial moment, when one offers one's own life to substitute for that of another who has been doomed, carries with it great redemptive power. "For the first time in his life, Gershon truly felt sorry for all of his wrongdoings. Heartbroken, he kneeled before the monster, and awaited his end." But in this case the autosacrifice is accompanied by that for which the Psalmist cries out and what the ceremony of *tashlikh* is supposed to entail: "The sacrifice of God is a broken and contrite heart" (Ps 51). In one cataclysmic moment, the sea returns all that Gershon has cast into it, and faced with the consequences, Gershon does more than try to bribe the thing by giving himself instead, a life for a life. He actually feels the foreign emotion of remorse, the first step of *t'shuvah*. Because he has undertaken only the annual religious observance of dumping his misdeeds into the sea without also correspondingly changing his heart, he now experiences the cumulative effect of that false cleansing. Death comes from the ocean, the place in which he had hoped to hide his sins forever: death comes not for him but for the children he conjured by the wonder rabbi's prayers and now loves more than life. Yet "[Death] never came. The monster rose into the air like a great cloud. Its scales melted into raindrops that fell like a summer shower, cleansing the sea."

In the last and greatest of the tale's many ironies, the scales of the
monster, the sins of Gershon that had polluted the sea, now cleanse
it, transformed by his true repentance into pure rainwater.

The Story of the Sea Glass

The sea's ultimate exposure of what is furtively thrown into it un-
derlies Anne Wescott Dodd's *The Story of the Sea Glass*, a children's
book about beach-combing and memory.[14] Red sea-polished glass is
the rarest kind to find among the many other pieces of blue, green,
white, or brown. Each piece reflects its origin as a bottle thrown into
the sea; red glass bottles are rare, and one would not jettison such a
treasure. In this story, the red sea-glass carries with it an indictment
of a long-ago trespass as well as a message of the impossibility of hiding secrets in the sea forever. It also signifies the sea's ability to effect
transformation, both physical and moral.

Nicole, who is about nine years old, begs her grandmother to take
her to see the coastal Maine island where Nana grew up and which
she so often mentions in stories she tells Nicole. They pay a visit, and
although Nana's childhood house has new owners, they can explore
the beach, filling their pockets with periwinkles and sea glass. When
both Nicole and Nana discover the small red sea-glass, "still wet
from the receding tide," Nana exclaims, "No, this can't be." She sits
Nicole on her knee: "You'll soon see why I have never told you this
story." She tells her granddaughter the tale of one restless summer
day many years ago, when Nana herself was Nicole's age. Wandering
from room to room, the girl was drawn to the beautiful red glass vase
that was her great-grandmother's most prized possession, a wedding
gift, sitting on the shelf of the equally forbidden formal parlor. Her
mother "must have reminded me a hundred times. 'That vase especially is off-limits to you.'

"But the vase was red—my favorite color. Like a very powerful
magnet, it pulled me right into the room. And, of course, I could

not resist picking it up." The girl carried it to the sunniest window, where looking through it, she saw the coastal world magically transformed into shades of rose and red. Then suddenly, something brushed her leg—her cat, Minou; she jumped and dropped the prized vase, shattering it into bits. Young Nana quickly and carefully gathered the sharp pieces and cradled them in her dress, tiptoed past her mother in the kitchen, "then sneaked by to get to the shore."

There she found a "moon tide," a tide much lower than usual, which forced her to walk very far out on the beach, "past the spot where we usually dumped the cracked canning jars and chipped china cups. I climbed over the slippery rocks to reach the water's edge. With a huge sigh of relief, I tossed the broken glass into the ocean."

The guilty child immediately thought to go to the beach with the broken vase and throw the pieces into the sea. Why she does this is revealed by the fact that the family routinely threw away broken household containers at the water's edge. Her action of doing the same with the red glass resolves her acute anxiety. Her "relief" mirrors the relief, expressed centuries earlier and in a very different story, of the gods and sages who watch the mare-fire enter Ocean, no longer a threat to the triple worlds. In the world of "thrown away," the sea is often the ultimate "away."

Of course the ruse did not work, and the girl was discovered and harshly punished. As she grew older, not only did guilt haunt her, but so did the memory of what was lost: "Whenever I see a red sun rising, or setting, or red leaves in the fall, I think, If only . . . if only I had saved just one piece of that special red glass." She did not retrieve a piece; one wonders whether because of the glass's associations with shame, with her breaking of family taboo, or, perhaps most of all, with lost beauty, perfection destroyed in an instant.

As her granddaughter Nicole holds up the red glass, now clouded by its decades in the saltwater, Nana remarks on the ocean's unexpected revelation.

I thought the sea would hide my mistake. But maybe the sea fooled me. You know, it's amazing how the sea can take a shard of jagged, broken glass and keep it for years, tossing and pounding it until it becomes sea glass, smooth and round. This red glass is frosted now, and I can't see through it. It can't give us the rose-colored world I saw through the magical red vase. Yet the sun makes it glow in a wonderful new way. Here, Nicole—hold it up and see for yourself.

"I thought the sea would hide my mistake. But maybe the sea fooled me." Not only has the sea not hidden Nana's mistake, but it has waited until the perfect time to reveal it: when she has at last returned to her seaside home with her own granddaughter, who is now hungry to know more of Nana's island childhood. The life cycle has come full circle, and it is time. By tossing up the red glass fragment it has kept for decades—and washing it with wave-water until it gleams—the sea insists that the story be told, creating a picture of that childhood that is no longer falsely idyllic. "It can't give us the rose-colored world I saw through the magical red vase." Rather, Nana's history is complexified. Nana says that she never before told Nicole the story, nor was she planning to, but now, confronted with the red shard, she must.

Finally, Nana allows the sea glass to become a symbol of transformation, its edges softened by time and the gradual agency of memory. "This red glass is frosted now. . . . Yet the sun makes it glow in a wonderful new way." What goes into the sea never comes out of it the same. Yet it can and some day will come out. Then the lens of perception must shift to see it rightly in its new condition.

The Girl in a Swing

An obscure work by a celebrated author, Richard Adams's *The Girl in a Swing* begins as an erotic novel but, through the gradual intervention of the sea as an inexorable moral agent, ends in horror.

A British antiques dealer on a business trip to Copenhagen, Alan Desland, meets and falls in love with Käthe Wasserman, a German secretary working and living in the city. At first she responds indifferently to his attraction but then urgently, passionately accepts his proposal of marriage. The couple arranges to settle back in England at his country home. All is idyllic there, as Käthe tries to integrate herself into Alan's world, his work, and his circle of friends; and their romance burns both hot and holy with a kind of mystic eroticism. But then, one thing at a time, things turn slowly queer: Käthe refuses a church wedding and avoids all churches; when at last persuaded to attend a Eucharist, she faints upon receiving the cup and spills its contents everywhere. Alan dreams repeatedly of diving underwater to discover a decomposing female corpse. Finally, after months of dreamlike confusion, the walls themselves drip water; a child is heard desperately weeping in the garden; and a phone call Alan places to a Copenhagen associate is intercepted by "gurgling, watery noises" and a little girl, speaking in German, who says, "Mummy? Mummy, I'm coming as fast as I can. . . . I'm coming—soon—only it's such a long way—"

In the novel's dénouement, Alan and Käthe drive in an effort to preserve their sanity to a lonely stretch of the sea near Sidmouth. They make love on the sand, where Käthe strips off everything, even her wedding ring: "an elegy."

> The level, still sea was moving, ripping unnaturally. Something was disturbing it, something was approaching the surface, though with difficulty, it seemed—something close at hand, not twenty feet from where we were lying. A higher wave, softly turbulent, flowed forward and round us, soaking my clothes and very cold upon my naked loins. The shock brought me to me myself and I knew once more that I was lying on the beach, holding Käthe in my arms. She had turned her head...staring, wide-eyed and unbreathing, at the water. Following her gaze, I saw the surface break and saw what came out of the sea.

What came out of the sea, groping blindly with arms and stumbling with legs to which grey, sodden flesh still clung, had once been a little girl.[15]

Alan loses his senses and runs, "sobbing with a terror as much like normal fear as a leopard is like a cat." He is found bleeding, naked, and terrified and is taken by the police as a rape suspect to the same hospital where Käthe was brought by a motorist, naked and out of her mind. She is dying, as it turns out, from an ectopic pregnancy that has ruptured her Fallopian tube. As Alan sees her beautiful face ravaged, he kisses her and turns from the bed, weeping. She whispers to him, "*Mögst du nicht Schmerz durch meinen Tod gewinnen. Ich hatte kein Mitleid.*"[16]

The pelvic exam undertaken when Käthe was brought in reveals that she has already born a child, something Alan had already sensed and repressed when he discovered the receipt for a toy tortoise purchased in Copenhagen in the seam of one of his wife's handbags. This was a very young daughter by a former relationship, of whose existence Alan had been unaware and which Käthe had kept hidden from him. A chance remark Alan made during their courtship in Denmark about his strong antipathy to having children caused Käthe to fear that she would lose her chance to marry him if she told him of her motherhood. And so, desperate, just before her departure to England to marry him, Käthe drove her child north of Helsingør and drowned her. This was the one for whom Käthe "had no pity."

The dead child rejects her rejection and refuses to remain quietly decomposing. Instead she struggles across the bottom of the sea to England. The little girl returns not as a guilty memory that haunts her mother, her murderer, but as a surreality: her sea-eaten body, animated by the love that once bound the two of them, is an objective correlative for that memory. The sea does not remain a passive receptacle of Käthe's crime by sequestering her drowned daughter. Instead it manifests itself as a powerful, relentless, moral agent by confronting the woman with her deed's pitiful evidence, itself undead. The sea deals Käthe the madness and death she has earned. The sea bears the symbolic agency of karmic retribution. The inescapability of our moral history is what makes *The Girl in a Swing* a novel of horror.

Thoughtlessly discarded sins; red-glass shards from a shattered vase thrown fearfully into the cold island tide of Maine; the drowned daughter of a woman horribly intent on beginning a new life at any cost: all of these come back from the sea, each bearing as much symbolic weight as a literal sea-change "into something rich and strange." Explicitly religious or not, these tales of the marine return of what had been hidden, cast into the sea, share the tenor of anxiety and temporary relief that characterize the great apocalyptic accounts we considered earlier. But in these stories, unlike the stories of Ocean's fire-mare or the resurrection of the dead, the return is unexpected and hence surely unwelcome. Gershon sinks to his knees in despair when the hideous serpent, made up of all of his many years of misdeeds, towers up out of the sea to devour what is now most precious to him. Upon seeing the red sea-glass, an emblem of a childhood crime and the attempt to cover it up for good, Nana says, "No, this can't be." And Käthe, lying on the beach and staring at the sea, loses her mind and soon her life when she sees her gray little girl, unrecognizable to anyone else, emerge from its depths.

"Here End the Works of the Sea, the Works of Love"

These three narratives, taken with their explicitly religious counterparts, reveal a new dimension of marine pollution—the kind that will not stay put, but that will resurface to confront. The fiction of irretrievability—that things are gone forever when the ocean swallows them—itself dissolves. Lacking the developed, self-referencing beliefs that characterized traditional human systems of religious pollution and purification, contemporary human societies have turned indiscriminately to the sea to wash away our offscourings. Every manner of toxin streams into marine waters, affecting every denizen from the top to the bottom of the food chain. Plastic floats on the surface of the deepest and most remote seas; nuclear waste lies at the ocean's bottom, where lithic plates grind away, plowing it under at the edges, but also potentially moving such waste toward undersea volcanoes, whose heat could set off disastrous explosions. It is perhaps inconceivable to us that the restless, ever-changing flood that rolls over more than two-thirds of the surface of our planet could itself be vulnerable, or that in its vulnerability lies our own.[17] The human illusion remains that the sea must surely always renew itself, even as it dilutes our carrion and makes it "disappear." But "thrown away" is shown to be a chimera: there is no "away."

The history of religion shows that we are not the first to have thought or acted this way. A paradox is embodied in contemporary attitudes toward the sea that strangely mingle awestruck love with relentless abuse. That abuse, I have suggested, is based in a chronic, albeit not universal, response to the natural qualities of the sea itself. The affection for the sea that William Faulkner expressed to his mother—the love that catches our throat when we see it—is also freighted with peril for the beloved. We love the sea because it has no limits—because it seems as though it can do anything, take anything from us of which we would be rid. Because of that belief and that affection, just as the gods and sages felt for the vast, humble Ocean, who pressed its palms together and bowed, accepting the doomsday mare for the sake of the inhabited world, we are slowly compromising it, perhaps past the point of no return.

The selfless quality of the sea that is stressed in the Hindu stories of the submarine mare resonates with the ecological critiques discussed in chapter 1, in particular Wolf's discussion of instrumental versus intrinsic value in thinking about ecosystems, and his proposed application of Kant's humanity imperative to them, whereby one must always treat a person as an end in itself, and never only as a means.[18] Even in the religious accounts where it is hypostasized or divinized, even where it has natural or moral agency, the ocean is treated as a means, not an end; how much more clearly is this "instrumental" attitude to the sea visible in the recent history of marine pollution. Going further, Nicholsen offers an interpretation of great environments such as the sea as scapegoat sacrifices whose violent destruction, à la Girard, allows for the relief of social anxieties and enables ordered culture-building and group cohesion.[19] But will the "works of the sea" come to an end? Perhaps there is also an unimaginable time when, at last, what we put into the sea might emerge from it for a final reckoning.

Just as Samudra voluntarily releases the mare-fire of doom at the end of time and the sea of Revelation gives up all the dead that were in it, so the World Ocean, with its tidal currents and patterns of waves, has begun to spit back what is foreign to it. We thus confront the nondissolving, ugly reminders of what we had discarded, the consequence of our ascription of an instrumental rather than an intrinsic value to the sea. The beaches are strewn with what was supposed to have been jettisoned and carried off—plastics, styrofoam, syringes, condoms, sewage, biomedical waste. These are also unexpected and unwelcome, like the reconstituted sins in the Hasidic tale of Gershon, the red shard in the sea-glass story, and the denouement of *The Girl in a Swing*. Here, then, both in the reality of our coasts and in the spheres of the religious and literary imaginations is the fulfillment of Marilyn Strathern's warning that "pollution surprises by its untoward nature, an unlooked for return."

I have tried in this book to argue that human habits of thought and action were, and still remain, a kind of ritualized, response to human constructions of the ocean's physical qualities. Such ideas about the nature of the sea may be culturally reinforced, but they are daily reinscribed by the testimony of our eyes. In tons of water, in saltiness, in bottomless depth and endless horizon, and, above all, in many forms of ceaseless motion, human populations, especially those who live along the littoral, see—and have always seen—in the world's oceans a mighty, efficacious means of "cleaning" our habitus and making it safe, clean, and viable. Our impressions have lent themselves over the years to chronic, unreflective marine pollution. Hence they are understandable but no longer realistic.

Assumptions about the ocean's ability to sequester and to purify, which have been expressed religiously in the traditional conflation of moral and religious danger—the ancient Greek catharsis of murder's stain in the sea; the Inuit migration of human transgressions to the hair of an angry undersea spirit; the medieval Hindu mare-fire, burning with ill-managed divine wrath, who must dwell on the submarine floor and feed on seawater—are now, in the developed world, expressed literally in the entrenched practices of shoreline pollution and ocean dumping. It is only relatively recently that these practices have come to be challenged, and one may well argue that it is only in the industrial age that they have been deleterious. Perhaps the persistent myth of the ancient cathartic ocean has come to the point where its dangers outweigh its lyrical beauty.

Perhaps the sea can no longer wash away all evils of humankind.

~~~~~~
Notes

Preface

1. The exception to this is Bali, where the sea has always been the abode of demons; it could not be more evil or more impure. In a powerful polarity of orientation, a Balinese person both sleeps and is buried with his or her head—the locus of cognition and enlightenment—in the direction of one of the sacred peaks of Gunung Agung, Mahameru, or Batur, the source of all purity—and his or her feet toward the dangerous, death-dealing sea, in effect renouncing or, better, counteracting it. Balinese observe the American attraction to the seashore with a degree of astonishment. In Bali it is traditionally believed that the place where one's afterbirth is buried always pulls one home. Because Americans do not bury afterbirths, but rather allow them, as medical waste, to be flushed into the sea, some Balinese believe that is why their American visitors feel compelled to be near their spiritual home (personal communication, Kristin Scheible, Assistant Professor of Eastern Religions, Bard College, April 12, 2006).

2. As cited in Amy Waldeman and David Rohde, "Fearing a Sea That Once Sustained, Then Killed," *New York Times,* January 5, 2005.

3. Ibid.

4. Ibid.

5. Ibid.

6. Ibid.

7. J.R.R. Tolkien, *The Lord of the Rings* (1955; repr. Boston: Houghton Mifflin, 1994), 941.

8. Dorrick Stow, *Oceans: An Illustrated Reference* (Chicago: The University of Chicago Press, 2005), 236.

9. In keeping with the agenda of the international Disaster Reduction Conference in Kobe, Japan, in January 2005, the United States is considering the global expansion of existing tsunami warning systems, consisting of buoys, wave gauges, and seismic sensors, including the one being designed by Australian scientists for the Indian Ocean.

1. The Dutch Bread Man: Ocean as Divinity and Scapegoat

1. Associated Press, Amsterdam, August 10, 1992.

2. Participants in ritual disputes often refer to one other's actions as either "mindless" or "meaningless"; "empty" is also applied to criticisms of particular rituals. Scholarly caricatures of ritual, albeit made in more nuanced ways, can be observed in Vedologist Frits Staal's *Rules Without Meaning: Ritual, Mantras, and the Human Sciences* (New York: Peter Lang, 1989). An opposite school sees ritual as having the extralinguistic potential most perfectly to express most perfectly a philosophical thought or theological complexity. For example, see Catherine Pickstock, *After Writing: On the Liturgical Consummation of Philosophy* (Oxford: Blackwell, 1998).

3. Dutch religious groups also protested the plan. Jan van Rossem, a Reformed Church minister, stated, "We strongly object to the idea of a sacrifice to the sea. . . . The Bible says sacrifices should be made only to God."

4. David Suzuki, with Amanda McConnell, *The Sacred Balance: Rediscovering Our Place in Nature* (Amherst, N.Y.: Prometheus Books, 1998), 53.

5. James Hamilton-Paterson, *Seven-Tenths: The Sea and Its Thresholds* (London: Hutchinson, 1992), 5.

6. William Faulkner, letter to his mother, Maud Faulkner, October 6, 1921. In William Faulkner, *Thinking of Home: William Faulkner's Letters to His Mother and Father, 1918–1925* (New York: Norton, 1991), p. 145. The misspelling of "can't" is in the original.

7. Ibid.

8. Catherine Keller, *Face of the Deep: A Theology of Becoming* (London: Routledge, 2003).

9. Mary Oliver, "The Sea," in *American Primitive: Poems* (Boston: Little, Brown, 1983), 69–70.

10. Aldo Leopold, *A Sand County Almanac* (New York: Oxford University Press, 1966), 251.

11. Clark Wolf, "Environmental Ethics and Marine Ecosystems: From a 'Land Ethic' to a 'Sea Ethic,'" in *Values at Sea: Ethics for the Marine Environment*, ed. Dorinda Dallmeyer (Athens: University of Georgia Press, 2003), 29–30.

12. Ibid., 24–25.

13. Adalberto Vallega, *Sustainable Ocean Governance: A Geographical Perspective* (London: Routledge, 2001), 213.

14. Ibid.

15. Ibid.

16. Shierry Weber Nicholsen, *The Love of Nature and the End of the World: The Unspoken Dimensions of Environmental Concern* (Cambridge: MIT Press, 2002), 153–154.

17. See Kevin Hetherington's persuasive analysis of consumption and disposal as carrying the broader implications of the management of absence in social relations, in his essay "Secondhandness: Consumption, Disposal, and Absent Presence," *Environment and Planning D: Society and Space* 22 (2004): 157–173. Hetherington's work parallels some of the structural linkage I suggest between anthropological principles and patterns of contemporary waste management.

18. This was emphatically theorized by Mary Douglas in *Purity and Danger.* "The whole universe is harnessed to men's attempts to force one another into good citizenship. Thus we find that certain moral values are upheld and certain social rules defined by beliefs in dangerous contagion, as when the glance or touch of an adulterer is held to bring illness to his neighbours or his children." See Mary Douglas, *Purity and Danger: An Analysis of the Concept of Pollution and Taboo* (1964; repr. with a new preface by the author, London: Routledge, 2004), 4.

19. Rob Shields, "Spatial Stress and Resistance: Social Meanings of Spatialization," in *Space and Social Theory: Interpreting Modernity and Postmodernity*, ed. Georges Benko and Ulf Strohmayer (Oxford: Blackwell, 1997), 187.

20. Hamilton-Paterson, *Seven-Tenths*, 5 (italics in original).

21. See the seminal analysis by Jonathan Z. Smith, "The Influence of Symbols on Social Change: A Place on Which to Stand," in *Map Is Not Territory: Studies in the History of Religions* (Chicago: University of Chicago Press, 1993). Smith writes about Resnick's 1918 *The Ocean in the Literature of the Western Semites*: "This locative vision of man and the cosmos is revealed in a variety of descriptions of the places in which men stand. The world is perceived as a bounded world; focussing on the etymological roots, the world is felt to be an *environ*ment, an *ambi*ance. That which is open, that which is boundless . . . is seen as the chaotic, the demonic, the threatening. . . .

The desert and the sea are all but interchangeable concrete symbols of the terrible, chaotic openness. They are enemy par excellence. To battle against the power of the waters a divine warrior is required: Baal versus Prince Sea (Zabul Yam), or the seven-headed water dragon, Lotan; Marduk versus Apsu and Tiamat; Yahweh against Leviathan, Rahab or the Sea (Yam); Nunurta against Kurti; victory establishes those two inseparable companions: divine kingship and cosmic order. Order is produced by walling, channeling, and confining the waters" (134).

22. The nineteenth-century and early-twentieth-century European and American view of the sea, the amphitheater of great sailing ships, as a place of escape from the ugliness, brutality, and complexity of civilization spawned generations of romantic sea writers. In addition to sea-poems like those by Byron, Keats, Masefield, and Oliver Wendell Holmes, the prose works of Melville, Conrad, and Richard Henry Dana, among many others, enshrined the sea. As Jonathan Raban writes, "To turn one's back on the overcivilized world by running away to sea in search of transcendent verities was a romantic (with a small *r*) gesture, performed so often that it became something dangerously close to an established social ritual. In the dreamscape in which such actions are plotted, the sea-life was natural, pure, free of the constraints of the factory and the city. Yet in reality, the sea itself was beginning to be invaded by industrial machinery." Steam-, turbine-, and diesel-driven ships all were invented before 1900. See Jonathan Raban, introduction to *The Oxford Book of the Sea*, ed. Jonathan Raban (New York: Oxford University Press, 1992), 17.

23. As documented in recent years by, among other synthetic studies, Richard Ellis in *The Empty Ocean: Plundering the World's Marine Life* (Washington, D.C.: Island Press/Shearwater, 2003).

24. Deborah Cramer, *Great Waters: An Atlantic Passage* (New York: Norton, 2001), 57. Marine scientist Adalberto Vallega expresses a similar view in *Sustainable Ocean Governance:* "Mankind's historical attitude to the oceans was based on the belief that they constituted an inexhaustible reservoir of resources, a view widely shared during the principle decades of modern society. At the same time, it was regarded as a crucial arena for the creation of a new world order within which expanding equity and rational use of resources could be achieved. . . . The perception of the ocean as a finite resource reservoir became prevalent only in the late 1980s when it was estimated that the exploitable, sustainable global potential in biological resources was about 100 million tons a year and that this threshold was likely to be reached in a short term" (229).

25. James Brown and Mahfuzuddin Ahmed, "Consumption and Trade of Fish," Briefing paper no. 3, Conference of the Institute for European Environmental Policy, "Sustainable EU Fisheries: Facing the Environmental Challenges," Brussels, November 8–9, 2004, 2.

26. Mircea Eliade, *Patterns in Comparative Religion,* trans. Rosemary Sheed (1958; repr., Lincoln: University of Nebraska Press, 1966).

27. Mircea Eliade, *Cosmos and History: The Myth of the Eternal Return* (1959; repr., New York: Garland, 1985).

28. See Albert Henrichs, "'Thou Shalt Not Kill a Tree': Greek, Manichaean and Indian Tales," *Bulletin of the American Society of Papyrologists* 16 (1979): 85–108.

29. In the *Cad Goddeu,* attributed to Taliesin (Welsh) and Cath Maige Tuired (Irish). The Tuatha Dé Dannan enchant stones and trees to do battle against the Fomorions. The "walking forests" of Celtic legend find later incarnation in Macbeth's aggressive forest of Dunsinore and in the roaming Ents, the primeval tree-shepherds of Tolkien's *The Two Towers.*

30. "Then with his supernatural power Sakka (the king of the Gods, from Sakra, 'powerful,' the usual epithet of Indra) created a fire of burning logs, and told the Bodhisattva. Then arose the hare from its form of *kusa* grass (the sacred grass used in Brahmanic ritual) and said, 'If any small insects are in my hair they should not be destroyed,' saying which he shook himself three times (so as to let them escape) and, offering his whole body as a free gift, leapt up, and, like a royal swan alighting on a lotus bed, threw himself in an ecstasy of joy into the burning fire." From *Jātaka Tales,* trans. various hands, ed. E. B. Cowell (Cambridge: Cambridge University Press, 1895–1907), vol. 3, no. 316.

31. And keep him from rivaling the sun (Nanahuatzin as Tonatiuh) with his brightness, so as to dim his glowing competition with the newly born fifth sun, Tonatiuh (in the Florentine Codex and Leyenda de los Soles); see Karl Taube, *Aztec and Maya Myths,* 42, 2nd. ed. (Austin: University of Austin Press, 1995), 42.

32. Mircea Eliade, *The Sacred and the Profane* (London: Harcourt Brace Jovanovich, 1959), 118.

33. Ibid., 119 (italics in original).

34. Allan W. Larson, "The Phenomenology of Mircea Eliade," in *Changing Worlds: The Meaning and End of Mircea Eliade,* ed. Bryan Rennie (Albany: State University of New York Press, 2001), 56.

35. Douglas, *Purity and Danger,* 5. See Douglas's more recent reflections on the Levitical basis of her ideas in her essay "Sacred Contagion," in *Reading Leviticus: A Conversation with Mary Douglas,* ed. John F. A. Sawyer, Journal for

the Study of the Old Testament Supplement Series 227 (Sheffield: Sheffield Academic Press, 1996): 86–106.

36. Douglas, *Purity and Danger*, 6.

37. Ibid., 51.

38. Mary Douglas, *"Purity and Danger* Revisited," lecture, Institution of Education, the University of London, May 12, 1980; published in *London Times Literary Supplement*, September 19, 1980, 1045–1046.

39. Mary Douglas and Aaron Wildavsky, *Risk and Culture: An Essay on the Selection of Technical and Environmental Dangers* (Berkeley: University of California Press, 1983).

40. Douglas, *"Purity and Danger* Revisited," 1045.

41. Douglas and Wildavsky, *Risk and Culture,* 73. See also the discussion in Richard Fardon, *Mary Douglas: An Intellectual Biography* (London: Routledge, 1999), 156–162.

42. Patriarch Bartholomew of Constantinople, cited in Colin Woodward, *Ocean's End: Travels Through Endangered Seas* (New York: Basic Books, 2000), 226. In his documentation of a long personal journey over what he calls "dying seas"—the most ecologically compromised of earth's oceans—Woodward paints a poignant picture of a delegation of Orthodox Christian leaders traveling aboard the 38,000-ton *Venezilos* as it completed a weeklong circumnavigation of the Black Sea: "Out onto the ship's deck came a procession of bearded men in dark robes and tall smokestack-shaped hats, followed by a crowd of passengers. Solemnly they gathered at the rail overlooking the dark, brooding surface of the dying sea. These were the leaders of the Orthodox Christian world, the religious successors of the Byzantine Empire that once encompassed all the shores of the Black Sea. Standing side by side were the Patriarchs of Romania, Bulgaria, and Georgia, surrounded by their retainers. In their midst stood the Ecumenical Patriarch Bartholomew, head of the "mother church" in Constantinople, the former Byzantine capital.

With great ceremony, Bartholomew stood at the rail and blessed the dying sea below. The Black Sea had gained an unexpected ally" (225).

This was a voyage with a mission: the joint invitation of Bartholomew and the commissioner of the European Union, Jacques Santer, drew four hundred environmentalists, scientists, and religious leaders from around the world, "not only to see the Black Sea's problems up close but to discuss how religion and science might forge a partnership to save its environment" (226).

43. Ibid., 226.

44. Marilyn Strathern, "The Aesthetics of Substance," in Marilyn Strathern, *Property, Substance and Effect: Anthropological Essays on Persons and Things* (London

and New Brunswick, N.J.: Athlone Press, 1999), 61; cited in Hetherington, "Secondhandness," 162.

45. Stow, *Oceans*, 215.

46. Michael Specter, "The World's Oceans Are Sending an S.O.S.," *New York Times*, May 3, 1992.

2. The Crisis of Modern Marine Pollution

1. Stow, *Oceans*, 222.

2. Thor Heyerdahl, *The Ra Expeditions* (Garden City, N.Y.: Doubleday, 1971), p. 209.

3. Stow, *Oceans*, 227.

4. David Suzuki, with Amanda McConnell, *The Sacred Balance: Rediscovering Our Place in Nature* (Amherst, N.Y.: Prometheus Books, 1998), 71.

5. This tendency of pollutants to "settle to the bottom" was the organizing idea behind marine pollution theory during the 1960s and 1970s, when hydrologist Williard Bascom wrote his influential *Waves and Beaches: The Dynamics of the Ocean Surface* (New York: Doubleday, 1964). The effect was to underestimate the threat of chronic contamination to marine ecosystems.

6. Suzuki, *The Sacred Balance*, p. 70.

7. David Helvarg, *Blue Frontier: Saving America's Living Seas* (New York: Freeman, 2001), 3.

8. See ibid. The cold Labrador Current and the cold circumpolar currents at both ends of the globe meet the warmer currents at various points (such as in southern Newfoundland, creating fog), where micronutrients also are rich.

9. Michael Stocker, marine ecologist, Lagunitas, California, personal communication, November 2004.

10. Christopher L. Sabine, et al., "The Oceanic Sink for Anthropogenic CO_2," *Science* 16, vol. 305, no. 5682 (July 2004): 367–371.

11. Helvarg, *Blue Frontier*, 3.

12. Christopher Sabine, "Accumulating Carbon Dioxide Levels Starting to Change the Chemistry of the World's Oceans," National Public Radio interview by Richard Harris, *Morning Edition*, July 16, 2004.

13. Richard A. Feely, et al., "The Impact of Anthropogenic CO_2 on the $CaCO_3$ System in the Oceans," *Science* 16 (July 2004): 362–366.

14. Joanie Kleypas, "Accumulating Carbon Dioxide Levels Starting to Change the Chemistry of the World's Oceans," National Public Radio interview by Richard Harris, *Morning Edition*, July 16, 2004.

15. Michael Stocker, "Fish, Mollusks, and Other Sea Animals' Use of Sound, and the Impact of Anthropogenic Noise in the Marine Acoustic Environment," unpublished paper, Earth Island Institute, 2002, 1.

16. Stocker, "Fish, Mollusks," 6, noting the research by J. R. Potter and M. A. Chitre, "Ambient Noise Imaging in Warm Shallow Seas; Second-Order Moment and Model-Based Imaging Algorithms," *Journal of the Acoustical Society of America* 106 (1999): 3201–3210.

17. This is most clearly expressed by the icon of the environmental movement, in the shape of a bright red, eight-side stop sign that reads *Don't Use the Sea as a WC.*

18. Edward D. Goldberg, "The Oceans as Waste Space," in Center for Ocean Management Studies, University of Rhode Island, *Impact of Marine Pollution on Society,* ed. Virginia K. Tippe and Dana R. Kester (South Hadley, Mass.: Bergin, 1982), 26.

19. Ibid.; also see Douglas Allchin, "The Poisoning of Minamata," SHiPs Teacher's Network, *http://www1.umn.edu/ships/ethics/minamata.htm.*

20. Pew Oceans Commission, *America's Living Oceans: Charting a Course for Sea Change* (Arlington, Va.: May 2003), executive summary, vi.

21. U.S. Commission on Ocean Policy, "An Ocean Blueprint for the 21st Century," report delivered to the President and Congress on September 20, 2004 (*http://www.oceancommission.gov/*), executive summary, li. On PCBs, see Richard Gwynn. *Way of the Sea: The Use and Abuse of the Ocean* (Bideford, Devon: Green Books, 1987), 111.

22. U.S. Commission on Ocean Policy, "An Ocean Blueprint," xxxiii.

23. John Cairns, "Waterway Recovery," *Water Spectrum* (fall 1978): 28.

24. U.S. Commission on Ocean Policy, "An Ocean Blueprint," xxxiii.

25. Pew Oceans Commission, *America's Living Oceans,* vi. The Pew report calls such data "signs that our interactions with the oceans are unsustainable. Our activities, from those that release pollutants into rivers and bays to the overfishing of the seas, are altering and threatening the structure and functioning of marine ecosystems—from which all marine life springs and upon which all living things, including humans, depend" (vii).

26. Cairns, "Waterway Recovery," 28.

27. Stow, *Oceans,* 222.

28. Nichols Lenssen, "The Ocean Blues," in *The World Watch Reader,* ed. Lester R. Brown (New York: Norton, 1991), 43. This does not include the annual shipboard incineration of 100,000 tons of hazardous wastes.

29. Colin Nickerson, "As Ecologists Meet, Perils to Wildlife Shift," *Boston Globe,* March 14, 1992. Thailand's rivers continue to pour agricultural chemicals

into the Gulf of Thailand. The nation of China, powered by thousands of coal-fired plants, is still "dumping millions of tons of garbage into rivers, lakes, and the seas."

30. Nickerson, "As Ecologists Meet." The comment cited was made by To-shifumi Sakata, who at that time was the director of the Center for Information Technology at Tokyo's Tōkai University.

31. Aleksandr V. Souvorov, *Marine Ecologonomics: The Ecology and Economics of Marine Natural Resources Management, Developments in Environmental Economics*, vol. 6 (Amsterdam: Elsevier, 1999), 3.

32. According to Souvorov, "Intense economic development of seas and oceans, accompanied by the increasing human impact on marine ecosystems, endangers the existence of marine ecosystems. The ocean can purify and assimilate a certain amount of waste without significant ecological deterioration. But the amount and diversity of toxic substances entering oceans and seas do not allow us (at the present level of scientific and technological progress) to evaluate all the economic and ecological consequences of marine environmental contamination" (*Marine Ecologonomics*, 3).

33. As the report's introduction states, "The concept that the marine environment can be effectively used for waste disposal . . . has not been validated in a comprehensive assessment with land and atmospheric options." See Commission on Physical Sciences, Mathematics, and Applications (CPSMA), *Disposal of Industrial and Domestic Wastes: Land and Sea Alternatives* (Washington, D.C.: National Academy Press, 1984), 1. I interviewed James McCarthy at the Harvard University Center for the Environment on September 9, 2005.

34. CPSMA, *Disposal,* 1.

35. Cairns, "Waterway Recovery," 28.

3. The Purifying Sea in the Religious Imagination: Supernatural Aspects of Natural Elements

1. Mary Douglas, *Natural Symbols: Explorations in Cosmology* (1970; revised ed., New York: Routledge, 1996), 111. Douglas further elaborates, "A social structure that requires a high degree of conscious control will find its style at a high level of formality in stern application of the purity rule, denigration of the organic process, and wariness toward experiences in which control of consciousness is lost" (111). Here Douglas seems to extrapolate collective behavior and attitudes to "organic processes" from apparently Freudian starting premises, perhaps the alleged anal-sadistic stage at which children seek control over bodily

excretions yet neurotically compensate for their continual production and perceived uncontrollability.

2. Counterexamples do exist, although significantly they are the exception rather than the rule. In the preface I mentioned Balinese marine taboos, enacted in mortuary ritual. Here in Boston, in 1995, nine yellow-robed Tibetan monks from the Gaden Shartse monastery in India, led by monk Geshe Cheme Tsering, attempted to purify Boston Harbor at Rowes Wharf. They did this by chanting, blowing loud *tung-chens* (bronze horns), and pouring a glass container of tea directly into the waters. Clearly the sea could not purify the monks; the monks had come to purify the sea. The story of "Gershon's Monster," discussed in chapter 7, plays with both sides of the coin: the sea as a traditional purifier versus the sea in need of the purification of rain, symbolizing repentance.

3. Kimberley C. Patton, "'He Who Sits in the Heavens Laughs': Recovering Animal Theology in the Abrahamic Traditions," *Harvard Theological Review* 93, no. 4 (2000): 401–434.

4. See Jon D. Levenson, *Creation and the Persistence of Evil: The Jewish Drama of Divine Omnipotence* (New York: Harper & Row, 1985).

5. For the complete gamut of historical theological responses to this connection, one of the prime examples in world religions of the stylization, spiritualization, and politicization of a symbol originally taken from nature, see Joan E. Taylor, "The Asherah, the Menorah, and the Sacred Tree," *Journal for the Study of the Old Testament* 66 (1995): 29–54; L. Yarden, *The Tree of Light: A Study of the Menorah: The Seven-Branched Lampstand* (Ithaca, N.Y.: Cornell University Press, 1971); and Shubert Spero, "The Menorah: A Study in Iconic Symbolism," *Tradition* 14, no. 3 (spring 1974): 86–93.

6. Evan Zuesse, "Ritual," in *The Encyclopedia of Religion*, vol. 12, ed. Mircea Eliade (New York: Macmillan, 1987), vol. 12, 405. For the application of this definition, see Evan Zuesse, *Ritual Cosmos: The Sanctification of Life in African Religions* (Athens: Ohio University Press, 1979).

7. See Diana L. Eck, *Banaras, City of Light* (1982; repr. New York: Columbia University Press, 1999), esp. chap. 5, "The River Ganges and the Great *Ghāts*," 211–251.

8. Among other sources, see D. M. Stafford, *Tangata Whenna: The World of the Maori* (Auckland: Reed Books, 1996).

9. As Tertullian has it; also see John Chrysostom, *Homilae in Johannem*, xxv: "Verily I say unto thee, Except a man be born of water and of the Spirit, he cannot enter into the Kingdom of God" (Jn. 3:5) in "It represents death and burial, life and resurrection . . . when we plunge our head into water as into a tomb, the

old man is immersed, wholly buried; when we come out of the water, the new man appears at that moment." (Homiliae in Johannem, xxv.)

10. Theodor Schwenk, *Water: The Element of Life*, trans. Marjorie Spock (New York: Anthroposophic Press, 1989), 8–9.

11. See Jacob Neusner, *Tractate Miqvaot*, in his *Halakhah: Encyclopedia of the Law of Judaism* (Leiden: Brill, 1999).

12. See Jacob Z. Lauterbuch, "Tashlik: A Study in Jewish Ceremonies," *Hebrew Union College Annual* 11 (1936): 207–340. Jon Levenson, Professor of Jewish studies at Harvard Divinity School, remarks that the Gaon of Vilna, a great rabbi and archenemy of early Hasidism during the eighteenth century, objected to *tashlikh* in that "it becomes an occasion for socialization and even frivolity, thus destroying its penitential purpose and the introspective, self-critical mood of the day" (personal communication, August 2004).

13. Hilkhot Gezelah ve-Avedah in the Mishneh Torah, "The Laws of Theft," 11:10; cf. Dt 22:3. In the modern Greek islands, as I have personally heard attested, a kind of antiarchaeology goes on. The discovery of a buried artifact or inscription often delays for as much as several years the construction of a new hotel or house while evaluation and salvage archaeology is undertaken, thus causing economic hardship for the owners of the property. It is not uncommon for them to row the newly unearthed object out from shore by night and drop it into the sea, "reburying" it for good.

14. See T. Roger Wilson and J. Dennis Burton, "Why Is the Sea Salty? What Controls the Composition of Ocean Water?" in *Understanding the Oceans: A Century of Ocean Exploration*, ed. Margaret Deacon, Tony Rice, and Colin Summerhayes (London: UCL Press, 2001), 251–260, esp. 251–253.

15. Diogenes Laertius, *Pythagoras*, trans. R. D. Hicks (Cambridge: Harvard University Press, 1958), vol. 8, p. 35.

16. See James E. Latham, *The Religious Symbolism of Salt, Théologie historique*, vol. 64 (Paris: Éditions Beauchesne, 1982). The ancient Hebrews had a "covenant of salt" (Nm 18:19, 2 Chr 13:5).

17. Peter Dronke and Ursula Dronke, "Growth of Literature: The Sea and the God of the Sea," H.M. Chadwick Memorial Lecture, Department of Anglo-Saxon, Norse and Celtic, University of Cambridge, 1998, 21 (citing Kuno Meyer, *Zeitschrift für celtische Philologie* 5 [1905]: 4967, st. 2).

18. Yann Martel, *Life of Pi* (New York: Harcourt, 2001), 215–216.

19. Thich Nhat Hanh, *Friends on the Path: Living Spiritual Communities* (Berkeley: Parallax Press, 2002).

20. Rūmī, *Dīwān*, no. 2443.

21. Manāqib al-Āfarin, chap. 3, p. 379, para. 311. Cited in Annemarie Schimmel, *And Muhammad Is His Messenger: The Veneration of the Prophet in Islamic Piety* (Chapel Hill: University of North Carolina Press, 1985), p. 293, n. 48: "And that huge tree is the blessed existence of Muhammad, and the branches of this tree are the ranks of the prophets and stations of the saints, and those big birds are their souls, and the different tunes they sing are the mysteries and secrets of their tongues."

22. Gabriel Yiannis, "Organization Nostalgia: Reflections on the Golden Age," in *Emotion in Organizations*, ed. Stephen Fineman (London: Sage, 1993), 118–141. I am indebted to E. Garry Grundy III for drawing my attention to the metaphorical use of the ocean in these Freudian constructions of both childhood views of parental love and religious experience. Following Melanie Klein and others, Yiannis later traces the roots of human anxiety to the initial shock of realizing that the pleasurable breast will not always be there, which "establishes the first core separation between the ego and the object. The experience of separation defines us as specifically different from the rest of the world, and is marked by the first feelings of anxiety." See Gabriel Yiannis, *Organizations in Depth: The Psychoanalysis of Organizations* (London: Sage, 1999), 22. Melanie Klein deploys an aquatic metaphor in describing this stage as representing the loss of "fusion": "Putting the whole matter in terms of the infant's relations with the mother, the infant splits its relations with the breast into two parts: one is a good breast, belonging to a perfect mother who loves it absolutely; the other is a bad breast, belonging to a malevolent, withholding, dangerous mother. In this way, the infant establishes the root of a view of the world which divides it into elements that are all good, filled with love, with which it may perfectly fuse (water), and elements that are all bad, filled with hate, seeking to destroy it." See Melanie Klein, "Notes on Some Schizoid Mechanisms," in *The Selected Writings of Melanie Klein*, ed. Juliet Mitchell (New York: Free Press, 1946), 180.

23. Robert Buchbaum, "Hiking to Georges Bank: What If You Could Walk Under the Gulf of Maine?" *Sanctuary* 41, no. 6 (2002): 3–9.

24. See D. Helvarg, *Blue Frontier*, 3.

25. John H. Mitchell, "The Go-Between," *Sanctuary* 41, no. 6 (2002): 2.

26. Catherine Keller, *Face of the Deep: A Theology of Becoming* (London: Routledge, 2003), 222 (italics in original).

27. See especially the work of Frank Moore Cross, *Canaanite Myth and Hebrew Epic: Essays in the History of the Religion of Israel* (Cambridge: Harvard University Press, 1973); John Day, *God's Conflict with the Dragon and the Sea: Echoes of a Canaanite Myth in the Old Testament* (Cambridge: Cambridge University Press, 1985); and Neil Forsyth, *The Old Enemy: Satan*

and the Combat Myth (Princeton: Princeton University Press, 1987). On the connection in the Hebrew Bible of God's kingship with his victory over the angry waters and the establishment of his temple, as in Psalm 93, see the commentary by Jon D. Levenson, "Exodus and Liberation," in Jon D. Levenson, *The Old Testament, the Hebrew Bible, and Historical Criticism: Jews and Christians in Biblical Studies* (Louisville: Westminster/John Knox, 1993), 148–149: Psalm 93 sings:

> The Ocean sounds, O Lord,
> The oceans sounds its thunder,
> The ocean sounds its pounding.
> Above the thunder of the mighty waters,
> more majestic than the breakers of the sea
> is the Lord, majestic on high. (Ps 93: 3–4)

28. Yet as Jonathan Smith points out in his "The Wobbling Pivot"—an essay evaluating Eliade's ideas about sacred time and space—the ocean, and waters in general, is not the same kind of chaos as the Greek Chaos. At the annual Babylonian Feast of the Water Drawing, "the 'stoppers' to the channels leading down to the subterranean dangerous waters were opened and the upper and lower waters were mingled while a general carnival atmosphere and sexual license prevailed in the Temple. . . . [I]t is essential only to emphasize that this was no descent into chaos so that chaos could be overcome through recreation; but rather, a recognition of the power, the life-giving power, of chaos for insuring vitality and fecundity for the year to come. My sense is that on this issue, Eliade's discussion of water and water symbolism would provide a valuable corrective to the negative or neutral interpretation of chaos." Jonathan Z. Smith, "The Wobbling Pivot," in his *Map Is Not Territory: Studies in the History of Religions* (Chicago: University of Chicago Press, 1993), 97–98.

29. Martel, *Life of Pi*, 215.

30. The tradition of the *nāga-rāja*s guarding one of the eight portions undersea is noted in one of the two later lists of relics added to the *Mahāparinibbāna-sutta* by Sri Lankan elders, according to the fifth-century C.E. commentator Buddhagosa, and then expanded on in the thirteenth-century Sri Lankan chronicle, the *Thūpavaṃsa*. See the discussion of this tradition and those of the theft of the relics from the *nāga*s in Kevin Trainor, *Relics, Ritual, and Representation in Buddhism: Rematerializing the Sri Lanka Theravada Tradition* (Cambridge: Cambridge University Press, 1997), 120–132.

31. *Odyssey* 4.365–570. Proteus assumes countless numbers of different shapes as Menelaos tries to force him to divulge the fate of Odysseus: "a great bearded lion," a serpent, a leopard, a great boar; "and he turned into fluid water, to a tree with towering branches, / but we held on stiffly to him with enduring spirit" (4.456–459).

32. Celtic tales depict adventures of the dead on their way to their final resting place. The abode of gods (perhaps having Greek sources) has many names, among them Magh Mor, "great plain," Magh Mell, "plant of delight," and Annwn, "Abyss."

33. "And the Lord was sorry that he had made humankind on the earth, and it grieved him to his heart. So the Lord said, 'I will blot out from the earth the human beings I have created—people together with animals and creeping things and birds of the air, for I am sorry that I have made them'" (Gn 6:6–7). The arbitrary nature of this watery "omnicide" is hard to bear. God knows his flood will drown guilty and innocent alike, as well as all creatures (who surely are not culpable in the crisis of evil, at least not in the same way as human beings are; can a "creeping thing" sin?) except for the pairs on Noah's ark. Among the many stories of the miraculous from the Indian Ocean disaster of 2004 is that of the virtual absence of animal corpses from among the thousands dead. Animals around the perimeter of the sea sensed the quake, the wave, or both, and fled to higher ground before the sea made landfall. In this new flood, the animals saved themselves while the people perished.

34. See Toshio Akima, "The Songs of the Dead: Poetry, Drama, and Ancient Death Rituals of Japan," *Journal of Asian Studies* 41 (May 1982): 488; Wada Kiyoshi and Ishihara Michihiro, eds., *Zuisho wakoku-den* [*Chronicles of the Sui Dynasty*] (Tokyo: Iwanami shoten, 1977), 73, cited in Akima, "The Songs of the Dead," 488.

35. See the article by Sarah Brueggemann, "Fire on the Water," *Coastal Living* (May 2006): 116–19.

36. Because the sea, the domain of Aegir and Ran, was his home.

37. Legends persist of "ships of the dead" that were set ablaze and launched, but the archaeological evidence confirms only that the dead, especially the élite warrior dead, were sometimes buried in boats. Ships were also set on top of actual graves or as memorials over empty graves. The Gokstad ship, 76 feet long, was built around 850 B.C.E. and was made of oak. Robbed out during the thousand years since its interment in blue clay, the Viking ship was believed to have been that of a king; it also contained the remains of twelve horses, six dogs, and a peacock. Norse mythology had several worlds of death, all of which were reached after a long journey. The Icelandic chronicler Snorri

Sturluson (1179–1241) reports that the wicked go to Hel and then to Niflheimr though deep and dark valleys by Nastrans, the shore of corpses.

As we read in the 922 C.E. account by Aḥmad ibn Faḍlān, an emissary from the caliph in Baghdad to the Rus who lived along the Volga. In a detailed description of a ship cremation, he reports that a woman, a volunteer, was brought forward, lifted three times over a door frame, where she cried out that she saw her father, her mother, her dead relatives, and her master in a green paradise. She was then strangled on the dead man's pyre by an old woman known as "the angel of death." Faḍlān says the Norsemen decried Islamic burial customs, in which the corpse was "eaten by worms," and said to him, "We burn him in a moment, so that he enters paradise at once. . . . His master, out of love for him, sent the wind to carry him off in an hour." Was the wind fanning the flames or sailing the ship? Quoted by Jaqut in an English translation in *Antiquity* 8 (1934): 58–62. Everything was consumed by a fire lit by nearest kinsman, and a mound was built over the site, crowned by a wooden monument.

38. Carl-Martin Edsman, "Boats," trans. Kjersti Board in *The Encyclopedia of Religion*, ed. Mircea Eliade (New York: Macmillan, 1987), 260.

39. Mircea Eliade, *The Sacred and the Profane: The Nature of Religion*, trans. Willard R. Trask (New York: Harcourt, Brace, and World 1959), 26. For recent critical analysis of Eliade's thought and legacy, see the collected essays in *Changing World: The Meaning and End of Mircea Eliade*, ed. Bryan Rennie (Albany: State University of New York Press, 2001); and Douglas Allen's thoughtful *Myth and Religion in Mircea Eliade* (New York: Routledge, 2002).

4. "The Sea Can Wash Away All Evils": Ancient Greece and the Cathartic Sea

1. My translation.

2. The lack of archaeological evidence of Minoan burials and a chronic iconographic emphasis on marine life suggest that for at least some periods in their history, the Minoans may have buried their dead at sea.

3. When launching a ship, the ancient custom was to sacrifice a horned animal and sprinkle the ship's prow with blood, a custom continued to this day with the beribboned bottle of champagne shattered against the prow on launching. The sacrificial animal's hide or horns were suspended from the projecting part of ship, whose descendants were the red strips of cloth known as *foinikídes* hung from the spars of Cycladic fighting-boats. The victims' heads were eventually replaced by carved wooden figures, "figure-heads." See Dietrich

Wachsmuth, *ΠΟΜΠΙΜΟΣ Ο ΔΑΙΜΟΝ: Untersuchung zu den antiken Sakral-handlungen bei Seereisen* (Berlin: Ernst-Reuter-Gesellschaft, 1967).

4. Those who had been initiated into the Samothracian mysteries were under the protection of these deities and no longer feared drowning at sea.

5. Jean-Pierre Vernant, "Greek Cosmogonic Myths," in *Mythologies*, ed. Yves Bonnefoy (Chicago: University of Chicago Press, 1991), 368.

6. See the treatment by Petros Themelis, "Greece and the Sea: The Mythical Aspect," in *Greece and the Sea*, ed. Angelos Delivorrias (Athens: Greek Ministry of Culture and Benaki Museum, 1987), 58–60.

7. Walter Burkert, commenting on *Iliad* 20:4–9, in his *Greek Religion* (Cambridge: Harvard University Press, 1987), 174.

8. See the discussion in Gregory Nagy, *The Best of the Achaeans: Concepts of the Hero in Archaic Greek Poetry* (Baltimore: Johns Hopkins University Press, 1979), 196.

9. Trans. Gregory Nagy.

10. Trans. Richmond Lattimore.

11. "*Ōmēstēs*," *Iliad* 24.82, also used frequently in Greek literature of Dionysos, whose cult, at least in myth, involved *sparagmós*, ritual sacrificial dismemberment, and the eating of raw meat like beasts. Dionysos himself, though a god, takes refuge in the sea when pursued by the murderous Lykourgos in *Iliad* 6:136.

12. John Boardman, *The Greeks Overseas: Their Early Colonies and Trade*, rev. ed. (London: Thames & Hudson, 1999), 166, n. 15, fig. 203; Gisela Richter, *Greek Art* (1959; repr., New York: Dutton, 1980), 296, fig. 411.

13. The personification and deification of Pontus, the Mediterranean Sea, were apparently shared by a number of the other cultures that ringed it. Herodotus's famous account of the Persian invasion of Greece in 480 B.C.E. tells of how the bridges of ships crossing the Hellespont (the first "pontoon") were broken up by rough seas. Xerxes responded by having the sea flogged, cursing it: "You salt and bitter stream, your master lays this punishment upon you for injuring him, who never injured you. But Xerxes the king will cross you, with or without your permission. No man sacrifices to you, and you deserve the neglect by hour acid and muddy waters." (*Histories* 7.34, trans. Aubrey de Sélincourt). At last ready to cross the bridges, Xerxes brought incense and at sunrise "poured wine into the sea out of a golden goblet, and with his face turned to the sun, prayed that no chance might prevent him from conquering Europe or turn him back before he reached its utmost limits. His prayer ended, he flung the cup into the Hellespont and with it a golden bowl and a Persian *acinaces*, or short sword. I cannot say for certain if he intended the things which he threw into

the water to be an offering to the sun-god; perhaps they were, or it may be that they were a gift to the Hellespont itself, to show he was sorry for having caused it to be lashed with whips" (*Histories* 7.54). See the discussion of such ancient Near Eastern ideas in Abraham Malamat, "The Sacred Sea," in *Sacred Space: Shrine, City, Land*, ed. Benjamin Kedar and R. J. Zwi Werblowsky (New York: New York University Press, 1998), 51.

14. For a discussion of the solar hero's recapitulation of the sun's death and rebirth in Okeanos, see Nagy, *The Best of the Achaeans*, 196.

15. *Odyssey* 11.134–136.

16. On Ino, see Samson Eitrem, *Realencyclopädie der classischen Altertumswissenschaft*, vol. 12, cols. 2293–2306; Walter Burkert, *Homo Necans: The Anthropology of Ancient Greek Sacrificial Ritual and Myth*, trans. Peter Bing (Berkeley: University of California Press, 1983), 178ff.; Jan N. Bremmer, "Ino-Leucothea," in *The Oxford Classical Dictionary*, 3rd ed., ed. Simon Hornblower and Anthony Spawforth (Oxford: Oxford University Press, 1996), 760. Ino jumped into the sea with her little son Palaimon-Melikertes, whose body was carried to the Isthmus of Corinth on the back of a dolphin, where he had mystery rites and a hero shrine, the remains of which are still visible. Ino was worshiped in "the whole of Greece" according to Cicero (*De natura deorum* 3.39), and her festivals, at places like Delos, Teos, and Miletus, seemed to have focused on initiation and forms of social reversal. Ino's veil, pregnant as it was with her marine apotheosis, was one of the many powerful, strange entities in the ancient Greek sea that offered salvation to the shipwrecked. St. Nicholas, the second-century patron saint of sailors from Asia Minor, seems to have assumed much of this complex array of functions when Christianity reached the coastal Mediterranean.

17. Burkert, *Greek Religion*, 172.

18. See as well the discussion of the homecoming of Odysseus vis-à-vis concepts of the sea in antiquity in Raimund Schulz, *Die Antike und das Meer* (Damstadt: Primus Verlag, 2005), 17–28.

19. Christos Boulitis, "The Aegean Area in Prehistoric Times: Cults and Beliefs About the Sea," in *Greece and the Sea*, ed. Angelos Delivorrias (Athens: Greek Ministry of Culture and Benaki Museum, 1987), 32–33.

20. *Etymologicum magnum* 127.13 (*phúsei tò húdōr tês thalássēs kathársion*).

21. Theophrastus, *Characters* 16.2, says of *this* character that "he would seem to be one of those who carefully sprinkle themselves at the sea." See the discussion in Robert Parker, *Miasma: Pollution and Purification in Early Greek Religion* (New York: Oxford University Press, 1990), 227 and 307. Compulsive ritual ablution in the sea is only one behavior spoofed in the character of the *deisidaímōn*, who, as Parker notes, not only continually washes his hands but

also sprinkles his body with lustral water, chews laurel, avoids scenes of birth and death as well as tombs, and seeks out the priests of the Orphic mysteries (*Orpheotēlestai*) every month.

22. *On the Sacred Disease* 148.44ff., 1.42 G. Text cited in Parker, *Miasma*, 229, and n. 30. Parker notes the absence in the ancient text of the crossroads, bailiwick of the triple Hekate, one of the most common places of disposal of *kathármata*.

23. Pausanias 2.17.1; see S. Eitrem, *Opferritus und Voropfer der Griechen und Römer* (Kristiana, 1915), 84. On one classical vase now at the Ashmolean Museum in Oxford, even Apollo himself washes his hands at a lustral basin.

24. As Lady Macbeth discovered, some stains cannot be washed away. So, too, the Second Messenger in Sophocles' *Oedipus the King* 1227–1230, trans. David Grene: "Phasis nor Ister cannot purge this house, / I think, with all their streams, such things / it hides, such evils shortly will bring forth / into the light, whether they will or not."

25. Aeschylus, *Eumenides* 448–452, trans. Richmond Lattimore (italics added). Through the homeopathic mechanism of sacrificial blood redeeming blood spilled, Orestes has been partially purified. But it is by running waters that he was ultimately absolved and washed clean. Compare Euripides, *Electra* 794 and *Hippolytus* 653. The "running waters" could be a combination of multiple springs. Compare Empedocles B 143, and Apollonius Rhodius, *Argonautica* 3.860. The Souda says that a murderer's clothes must be washed in fourteen springs. See Parker, *Miasma*, 226 and n. 107.

26. Parker, *Miasma*, 227.

27. We know that an artificial salt-pool existed at the Erechtheion at the Acropolis, representing Poseidon's (losing) gift to the polis. In the struggle for control of Attica, Athena donated an olive tree and won dominion.

28. See the scholion to Aristophanes, *Acharnians* 747. Each *mústēs* bathed together in the sea with the pig, which each candidate had to sacrifice the next day "on his [or her] own behalf" to the goddesses. Ironically, today the Bay of Eleusis, 14 miles from Athens, is the site of no fewer than six hundred factories and plants of different sizes and functions, most of which directly pollute the air and pour toxic waste into the waters of the bay with almost no control. Greek officials in both government and private industry have been repeatedly warned over the past four decades about the dangers of their modus operandi, which amounts to the short-sighted ruin of the natural resource that is most precious to two of Greece's main industries: tourism and fishing. Because of Greece's economic woes, headlong policy of industrialization, and Byzantine political factionalism, meaningful industrial pollution controls have made slow progress.

29. Plutarch, *Life of Alibiades* 34.4; Xenophon, *Hellenica* 1.4.12. See also Walter Burkert, "Buzyge und Palladion: Gewalt und Gericht in altgriechischem Ritual." *Zeitschrift für Religions- und Geistesgeschichte* 22 (1970): 356–368. George Garrett, "Statue Bathing and the Plynteria Festival," paper delivered at the Annual Meeting of the American Philological Association, Washinton, D.C., December 28, 1993.

30. Franciszek Sokolowski, *Lois sacrées des cités grecques* (*LSCG*) 154 B 17–32 (Paris: Éditions de Boccard, 1969).

31. *Inscriptiones Graecae* (*IG*) II2 1006.11–12, 75–76; *IG* II2 1108.9–10; *IG* II2 1101.10–11.

32. Pausanias, *Description of Greece*, trans. W. H. S. Jones (Cambridge: Harvard University Press, 1918–1935). I am indebted to Robyn Walsh, Brown University, for drawing this account in Pausanias to my attention.

33. Pausanias, *Description of Greece*, trans. Peter Levi (Harmondsworth: Penguin, 1971), 316.

34. See, among others, Albert Henrichs, "The 'Sobriety' of Oedipus: Sophocles OC 100 Misunderstood," *Harvard Studies in Classical Philology* 87 (1983): 87–100; Walter Burkert, "Opferritual bei Sophokles: Pragmatik–Symbolik–Theater," *Der altsprachliche Unterricht* 18, no. 2 (1985): 5–20; Hugh Lloyd-Jones, "Ritual and Tragedy," in *Ansichten griechischer Rituale. Geburtstags-Symposium für Walter Burkert*, ed. Fritz Graf (Stuttgart: Teubner, 1998), 271–295; and P. E. Easterling, "Tragedy and Ritual," in *Theater and Society in the Ancient World*, ed. Ruth Scodel (Ann Arbor: University of Michigan Press, 1993), 7–24. For the most recent treatment of drama as itself a ritual idiom, see Barbara Kowalzig, "From Drama to Ritual: Performances of, by, and for Dionysos" and Kimberley C. Patton, "From Ritual to Drama: A Comparative Analysis," both in *From Ritual to Theater: The Prehistory of Theatre in Ancient Greece and Elsewhere*, ed. Margaret Miller and Eric Csapo (forthcoming, Cambridge: Cambridge University Press, 2007).

35. *Oedipus the King*, trans. David Grene, in The Complete Greek Tragedies, *Sophocles I*, ed. David Grene and Richmond Lattimore (Chicago: University of Chicago Press, 1954). Grene's translation is used unless otherwise noted.

36. As Parker points out, while the verb *ekpémpō*, "send out," usually refers to human exiles, it can also be sometimes used of polluted remains (*kathármata*), "as though there were something slightly animate about them" (Parker, *Miasma*, 230).

37. Parker, *Miasma*, n. 133, citing, among other sources, Aeschylus, *The Libation Bearers* 98 and Euripides, *Andromache* 193.

38. Parker, *Miasma*. Also cited: *Orphic Hymns*. 11.23, 14.14, 13.16, 17.11. Otto Weinreich studied the phenomenon of *apopompé* in ancient Greek religion.

39. This dyadic opposition appears much later in a medieval Greek Orthodox icon, a fresco from an apse in a medieval Greek church in Cyprus, which represents Land and Sea gazing at each other as hypostasized rulers. Land, the realm of what is ordered and redeemed, glares at the recalcitrant Sea, in whose waters ships are tossed while an octopus grins gleefully from the depths.

40. Sophocles, *Ajax*, trans. John Moore, in The Complete Greek Tragedies, *Sophocles II*, ed. David Grene and Richmond Lattimore (Chicago: University of Chicago Press, 1957). I have used Moore's translation unless otherwise noted. Greek text: Sophocles, *Ajax*, ed. W. B. Stanford (1963; repr., Bristol: Bristol Classical Press, 1981).

41. Euripides, *Iphigenia in Tauris*, ed. Maurice Platnauer (1938; repr., Bristol: Bristol Classical Press, 1984). *Iphigenia in Tauris*, trans. Witter Bynner, in The Complete Greek Tragedies, *Euripides II*, ed. David Grene and Richmond Lattimore (Chicago: University of Chicago Press, 1956).

42. The verb has the overt meaning "to purify" or "ritually cleanse," but it can also mean "to offer as a sacrifice." Both meanings are clearly implicated in Iphigenia's words to Thoas.

43. The translation of this passage is my own, since here, uncharacteristically, Bynner's distorts some crucial elements, mostly having to do with cult.

5. "The Great Woman Down There": Sedna and Ritual Pollution in Inuit Seascapes

1. Michael Brown wrote at the time, "Once the orbit of 2003 VB12 is known well enough (probably 1 year), we will recommend to the IAU [International Astronomical Union] Committee on Small Body Nomenclature—which is responsible for solar system names—that it be permanently called Sedna. Our newly discovered object is the coldest most distant place known in the solar system, so we feel it is appropriate to name it in honor of Sedna, the Inuit goddess of the sea, who is thought to live at the bottom of the frigid arctic ocean. We will furthermore suggest to the IAU than newly discovered objects in this inner Oort cloud all be names after entities in arctic mythologies" (*http://www.gps.caltech.edu/~mbrown/sedna/*).

2. Daniel Merkur observes, "Traditional Inuit religion is variously extinct, obsolescent, and persisting, depending on which Inuit band is under discussion." See Daniel Merkur, *Powers Which We Do Not Know: The Gods and Spirits of the Inuit* (Moscow: University of Idaho Press, 1991), vii. Modern Inuit artistic representations of Sedna abound, especially small sculptural form of wood, stone, bone, or ivory. One impressive example, by Arctic Bay artist Manasie

Akpaliapik, captures the moment when Sedna is entreated (and threatened) as she is brought up from the sea: *Shaman Summoning Taleelayuk to Release Animals* (Winnipeg Art Gallery, whalebone and narwhal ivory, 43.7 x 27.8 cm, 1989). See David Pelly, *Sacred Hunt: A Portrait of the Relationship Between Seals and Inuit* (Vancouver: Greystone Books and Seattle: University of Washington Press, 2001), 11.

3. See the discussions in Anne Fienup-Rirordan, *Agayuliyararput: Kegginaqutm Kangiit-llu* [*Our Way of Making Prayer: Yup'ik Masks and the Stories They Tell*] (Seattle: Anchorage Museum of History and Art in Association with University of Washington Press, 1996), 194; and Nelda Swinton, *La Déese inuite de la mer / The Inuit Sea Goddess* (Montreal: Musée des beaux-arts de Montréal, 1980), 15.

4. Narelle Bouthillier, "She Down There: Sedna and Her Sisters of the Deep," unpublished paper, Harvard Divinity School, 2004, 4.

5. For an emic challenge to the accuracy of the term *goddess* as applied to Sedna, see the six-part series by Rachel Attituq Qitsualik, an Inuit linguist, writer, and storyteller, entitled "The Problem with Sedna," *Nunatsiaq News*, March/April 1999. Qitsualik argues that because of Sedna's origins and because she does not inspire "worship" or reverence, but only dread, she is better understood as an "angry spirit." I would question whether either of these English terms do justice to the term *inua*, which emphatically describes Sedna. See Merkur's discussion cited in note 7.

6. Knud Rasmussen, *Report of the Fifth Thule Expedition 1921–1924: Intellectual Culture of the Igulik Eskimo*, vol. 7, part 2 (1929; repr., New York, AMS Press), 62.

7. Merkur, *Powers Which We Do Not Know*, 34. Merkur delineates the nature of particular *inue* by referring to their effects on their chosen location: "Outside the human mind, *inue* are specific in location to the phenomena whose forms they impart. Like the phenomena, they may variously be unchanging, mutable, or destructible. . . . *Inue* are completely autonomous and disinterested in people. Inuit can hurt themselves by abusing *inue* or derive benefits by being in accord with them. In both cases, the *inue* are what they are, with neither positive nor negative ambitions toward human beings. Because the Wind Indweller has a stern personality, Arctic weather is often fierce. In summer, his temper is better. Because the Sea Mother is jealous and vindictive, the sea is dangerous and miserly in its provision of game. Because the Moon Man has a benevolent disposition, the moon casts a benign light during the long winter nights" (33).

8. Ibid., 31.

9. Ibid., 32.

10. Ibid., 33.

11. Franz Boas, *The Central Eskimo* (1888; repr., Lincoln: University of Nebraska Press, 1964); Knud Rasmussen, *Intellectual Culture of the Igulik Eskimos*, vol. 7, nos. 1 and 2; *The Netsilik Eskimos: Social Life and Spiritual Culture*, vol. 8, nos. 1 and 2; and *Intellectual Culture of the Copper Eskimos*, vol. 9 of *Report of the Fifth Thule Expedition* 1921–1924 (1929, 1930, 1931, 1932; repr., New York: AMS Press, 1976); Diamond Jenness, *The Life of the Copper Eskimos*, vol. 12 of *Report of the Canadian Arctic Expedition* 1913–18 (Ottawa: F. C. Acland, 1922); Inge Kleivan, "Mitartut: Vestiges of the Eskimo Sea-Woman Cult in West Greenland," *Meddelelser om Grønland* 161, no. 5 (1960); Åke Hultkrantz, *The Religions of the American Indians*, trans. Monica Setterwell (Berkeley: University of California Press, 1979); Åke Hultkrantz, "The Religion of the Goddess in North America," in *The Book of the Goddess, Past and Present: An Introduction to Her Religion*, ed. Carl Olson (New York, Crossroad, 1983); Daniel Merkur, *Becoming Half-Hidden: Shamanism and Initiation Among the Inuit* (Stockholm: Amqvist and Wiksell, 1985); and Daniel Merkur, *Powers Which We Do Not Know.*

12. Swinton, *La Déese inuite de la mer*, 13.

13. The Netsilik "mother of the sea beasts" legend tells of the sacrifice of a girl, Nuliajuk, but she is very young. An orphan, she is thrown overboard from a raft when the entire settlement leaves Qingmertoq to find new hunting places. The motif of her desperately clutching the raft produces the same heartless response, chopping off her fingers, "and as she sank to the bottom the stumps of her fingers became alive in the water and bobbed around the raft like seals. That was the origin of seals. But Nuliajuk herself sank to the bottom of the sea. There she became a spirit, the sea spirit, and she became the mother of the sea beasts, because seals had been born of her fingers. She became mistress of everything else alive; the land beasts, too, which mankind had to hunt.

"So she had great power over mankind, who had despised her and thrown into the sea. She became the most feared of all spirits, the most powerful, and she controlled the destiny of men. Many taboos are directed toward Nuliajuk, though only in the dark period when the sun is low and it is cold and windy on earth—then life is most dangerous to live.

"Nuliajuk lives in a house on the bed of the sea. At the bottom of the sea there are lands just as on the earth, and she lives in a house that is built like a human house.

"Nuliajuk lives like a hermit, quick to anger, terrible in the ways she punishes mankind. She notices every little breach of taboo, she knows everything. Whenever people break a taboo toward her, she hides all the animals, for one thing, she shuts up all the seal in a drip-basin she keeps under her lamp. Then mankind begins to starve. People then have to call on shamans to help. Now, some shamans let their helping spirits do all the work, and the shamans themselves remain in their houses, summoning and conjuring in a trance. Some shamans rush at Nuliajuk to fight her. And there are some who draw Nuliajuk herself up to the surface of the land. They do it this way: they make a hook fast to the end of a long sealskin thong and throw it out of the entrance passage of a special house; the helping spirit sets the hook deep in Nuliajuk, and the shaman hauls her up into the passage. There everybody can hear her speaking. But the entrance from the passage into the living room must be closed with a block of snow, and Nuliajuk will keep trying to break it into pieces, in order to get into the house and frighten everybody to death. And there is great fear in the house. But the shaman watches the block of snow so that Nuliajuk doesn't get in. Only when she has promised the shaman to release all the seals into the sea again does the shaman take her off the hook and let her go free.

In that way a shaman, who is only a human being, can subdue Nuliajuk and save people from hunger and misery. . . . This is all we know of Nuliajuk, the sea spirit. She gave us seals, but she'd like to get rid of us, too." From "The Mother of the Sea Beasts," in *Traditional Stories of Eskimo and Indian People,* ed. Howard Norman (New York: Pantheon Books, 1990), 212–214.

14. See "The Sea Mother in Mythology," in Merkur, *Powers Which We Do Not Know,* 125–136.

15. Harold Seidelman and James Turner, "Sedna and the Shaman's Journey," in *The Inuit Imagination: Arctic Myth and Sculpture* (New York: Thames & Hudson, 1994), 76–78.

16. For example, the agonized *Sedna with a Hairbrush* by Igloolik sculptor Natar Ungalaq (stone, fur, and bone, 18.0 x 21.5 x 20.0 cm; National Gallery of Canada 3256.1–2, 1985); Sedna's wild hair is made of a black shock of fur. Some contemporary Inuit artists portray Sedna with fingers, in an important variance from the myth. Swinton speculates that this reflects a recognition of "the symbolic significance of the fingers as the primary link to the creation of the multitude of sea animals" (Swinton, *La Déese inuite de la mer,* 15).

17. Frank Boas, "The Eskimo of Baffin Land and Hudson Bay," *Bulletin of the American Museum of Natural History* 15 (1901): 130.

18. Both ceremonies are collectively enacted in the *qagge* (the "singing" or "dancing" house, a large *iglu* dedicated to that purpose).

19. Inge Kleivan, "Sedna," in *The Encyclopaedia of Religion,* ed. Mircea Eliade (New York: Macmillan, 1987), vol. 13, 166.

20. Boas, "The Eskimo of Baffin Land and Hudson Bay," 122.

21. Ibid., 120.

22. Swinton, *La Déese inuite de la mer*, 18. My translation.

23. One of the best ethnographic treatments of this complex topic remains that of Regitze Margrethe Søby, "The Eskimo Animal Cult," in *Essays Presented to Erik Holtved on His Seventieth Birthday, Folk (Dansk Etnografisk Tidsskrift)* 11/12 (1969/70): 43–78.

24. Cited in Pelly, *Sacred Hunt*, 11.

25. Rasmussen, *Intellectual Culture of the Igulik Eskimo*, 6.

26. Rasmussen, *The Netsilik Eskimos*, 242.

27. Boas, "The Eskimo of Baffin Land and Hudson Bay," 121–122.

28. Ibid., p. 121; cited and discussed in Swinton, *La Déese inuite de la mer*, 19.

29. Rasmussen, *The Netsilik Eskimos*, 225 (italics added).

30. For example, "In the beginning animals did communicate with people. It was told that in the beginning of first man, every living creature spoke, including humans, caribou, lemmings, mosquitoes." George Kuptana, b. 1914, Bathhurst Inlet, Nunavit; trans. John Nengoak, collected by Doug Stern, and cited in Pelly, *Sacred Hunt*, 12. Pelly comments on Kuptana, "Like all people and animals long ago, Inuit and seals spoke the same language and lived in absolute harmony, as equal partners in the natural world. Animals could also reason and react to events in much the same way as humans, and they were capable of human-like emotions. When it suited them, hunters could turn themselves into animals, and animals could become human beings. So too a seal could become a caribou or a fox. All creatures were equal in this regard, including humans."

31. Merkur, *Powers Which We Do Not Know*, 34.

32. Rasmussen, *The Netsilik Eskimos*, 225.

33. On Baffin Island, one of the epicenters of the Sedna myth cycle, a ritual was held annually in the autumn when storms broke open the ice and prevented the successful hunting of seals. This ritual, held in darkness, apparently *routinized* the intercessory response to accumulated transgressions and the cleansing thereof. It included a strange assault on the goddess herself, recapitulating the violence wrought on her during her human history as well as the antagonism that has characterized her relationships with human beings ever since her exile to the bottom of the sea. Inge Kleivan describes it as follows: "Sedna was harpooned through a coiled thong on the floor, which represented a seal's breathing hole. A shaman followed her and stabbed her, thereby cleansing her of the transgressions of taboos that had taken place the previous year (and thereby securing that she no longer would withhold the sea animals). When the lamps were lit again after the séance, blood was seen on the harpoon point

and the knife; the blood was an omen of good hunting in the future" (Kleivan, "Sedna," 166).

34. Rasmussen, *Intellectual Culture of the Igulik Eskimos*, 126.

35. Merkur, *Powers Which We Do Not Know*, 114.

36. Ibid.

37. Ibid., 127.

38. According to the Copper Inuit, she is suffering, like the people above, with no blubber for her lamp.

39. Boas, "The Eskimo of Baffin Land and Hudson Bay," 358.

40. Henry Rink, *Tales and Traditions of the Eskimo* (Edinburgh: William Blackwood and Sons, 1875), 39–40.

41. Rasmussen, *Intellectual Culture of the Iglulik Eskimos*, 127 and 173; Rink, *Tales and Traditions of the Eskimo*, 326.

42. Boas, "The Eskimo of Baffin Land and Hudson Bay," 497.

43. See Jarich G. Oosten, "The Symbolism of the Body in Inuit Culture," in *Visible Religion I: Commemorative Figures*, ed. H. G. Kippenberg (Leiden: Brill, 1982), 103–104.

44. See John Murdoch, "Ethnological Results of the Point Barrow Expedition," Bureau of American Ethnology, *Annual Report* 9 (1892): 3–441; Diamond Jenness, *The Life of the Copper Eskimos*, 186; Nicholas Gubser, *The Nunamiut Eskimos: Hunters of Caribou* (New Haven: Yale University Press, 1965).

45. Merkur, *Powers Which We Do Not Know*, 118.

46. Fienup-Rirordan; *Agayuliyararput*, 194.

47. Ibid.

48. See the discussion of hands and marine mammal flippers as representative of a working "meat-sharing partnership" in Canadian Arctic symbolic thought in Birgitte Sonne, "The Acculturative Role of Sea Woman," *Meddeleser om Grønland, Man and Society* 13 (1990): 9. Fienup-Riordan additionally records the ritual attention paid by the Yup'ik to the flippers of sea animals killed in the hunt and brought back home: they were "anointed" with fresh or sweet water, which as creatures of the salt sea, the animals were believed to crave, as mentioned earlier in this chapter (Fienup-Rirordan; *Agayuliyararput*, 194). In Yup'ik masks, other animal parts (feathers, flippers) adorning the mask are represented realistically, but the human-faced *inua* (human spirit) of the animal, reaching out from the center of the round mask of transformation or from behind the small in-built doors, has correspondingly human hands.

49. Rasmussen, *Intellectual Culture of the Iglulik Eskimos*, 127; and A. L. Kroeber, "Tales of the Smith Sound Eskimo," *Journal of American Folk-lore* 12 (1899): 306.

50. Again, the analogies to the confessional atmosphere of the hairdresser's chair are striking.

51. Boas, *The Central Eskimo*, 177–178. Rasmussen's report on the Copper Eskimos stresses smooth, clean hair instead. Henry Rink's seminal account of the Greenland version has the shaman either braiding it or tying it in a hair knot, as do polar shamans (*Tales and Traditions of the Eskimo*, 326); Erik Holtved, *The Polar Eskimos: Language and Folklore* (Copenhagen: Meddelelser om Grønland, 1951), vol. 152, nos. 1 and 2, 23. Also see the discussion in Swinton, *La Déese inuite de la mer*, 15, nn. 1 and 2.

52. Holtved, *The Polar Eskimos*, 23.

53. More violent variants on this systematic cross-examination existed: the shamans of the Fury and Hecla straits on Baffin Island would, after the questioning, recapitulate the myth of the sea animals' origins by newly breaking off Sedna's fingers: her nails, torn off, released the bears; her first and second finger joints, the seals; her knuckles, herds of walrus, and the lower part of her metacarpal bones, the whales. Iglulik shamans did the same. See Boas, *The Central Eskimo*, 177–178.

54. Merkur, *Powers Which We Do Not Know*, 114.

55. Swinton, *La Déese inuite de la mer*, 22.

56. Knud Rasmussen, *Across Arctic America: Narrative of the Fifth Thule Expedition* (1927; repr., Fairbanks, Alaska: University of Alaska Press, 1999).

6. "O Ocean, I Ask You to Be Merciful": The Hindu Submarine Mare-Fire

1. Wendy Doniger O'Flaherty [Doniger], *Women, Androgynes, and Other Mythical Beasts* (Chicago: University of Chicago Press, 1980), 231.

2. On issues of dating and content, see the commentary of Ludo Rocher, *The Purāṇas*, vol. 2, part 3, of *A History of Indian Literature*, ed. Jan Gonda (Wiesbaden: O. Harrassowitz, 1986), 222–228; and Wendy Doniger O'Flaherty, introduction to *Hindu Myths: A Sourcebook Translated from the Sanskrit* (Harmondsworth: Penguin, 1975), 14–18. Based on textual dependencies, Rajendra Chandra Hazra has argued that the voluminous *Rudra-Saā* of the *Śiva Purāṇa*, in which the story of the submarine mare appears, cannot be dated earlier than the fourteenth century C.E. See his authoritative "Problems Relating to *Śivapurāṇa*" *Our Heritage, Bulletin of the Department of Postgraduate Training and Research* (Sanskrit College, Calcutta) 1, no. 1 (1953): 65. On the "popularizing" nature of the Purāṇas and scholarly tensions regarding the quality of its Sanskrit and incorporation of multiple and self-contradictory sources,

including folk traditions, despite its own internal claims of "unified meaning" and "conformity with *Śruti*, the Veda," see Velcheru Narayana Rao, "Purāṇa as Brahmanic Ideology," in *Purāṇa Perennis: Reciprocity and Transformation in Hindu and Jaina Texts*, ed. Wendy Doniger (Albany: State University of New York Press, 1993), 97.

3. From *The Śiva Purāṇa*, trans. a board of scholars (Delhi: Motilal Banarsidass, 1973–1974).

4. Jaan Puhvel, *Comparative Mythology* (Baltimore: Johns Hopkins University Press, 1987), 277, as well as his wide-ranging discussion in this chapter (16), "Fire and Water," 277–283. A detailed study of water in Hindu mythology can be found in Frans Baartmans, *Āpaḥ, the Sacred Waters: An Analysis of a Primordial Symbol in Hindu Myths* (Delhi: B.R. Publishing, 1990). Naturalist Christopher Leahy raises the possibility that witnessing the eruption of submarine volcanoes was the inspiration for the fire-in-the-sea myths. (Personal correspondence, August 2005.) Such volcanoes continue constantly to create "fire contained by ocean." Perhaps the most impressive today is the new volcanic cone growing 2,300 feet beneath the Pacific off the Samoan archipelago. Named "Nafanua," the cone of sea-cooled magma has grown 1,000 feet in less than four years out of the summit crater of Vailulu'u Seamount, "an unpredictable and very active underwater volcano presenting a long-term volcanic hazard." Hubert Staudigel, et al., "Vailulu'u Seamount, Samoa: Life and death on an active submarine volcano," *Proceedings of the National Academy of Sciences* 103: 17 (April 25, 2006): 6448. Growing at its present rate, the volcano could reach the surface in a matter of decades. The hydrothermal vents near Nafanua's summit feed eels with shrimp carried their way by volcanic currents; within the cone, the same currents have created a toxic cavity, a "moat of death" where the decaying carcasses of fish, crustaceans, and squid killed by these hydrothermal emissions float in high concentrations.

5. For a comprehensive discussion of the mythical themes in this story, see Wendy Doniger O'Flaherty, "The Submarine Mare in the Mythology of Śiva," *Journal of the Royal Asiatic Society of Great Britain and Ireland* 1 (1971): 9–27. I make no attempt to reiterate or even to summarize Doniger's work therein on the Indo-European themes of horse, mare, fertility and destruction; my interest here is confined to the unique role of Ocean in the myth.

6. See Wendy Doniger O'Flaherty, *The Rig Veda* (Harmondsworth: Penguin, 1981), 104–107. In the *Avesta*, he is Apam Napāt (see Puhvel for the implications of the Indo-European etymology in *Comparative Mythology*, 277). Puhvel architecturally relates this and the Avestan version of the story to the Irish *Dindshenchas* story of Nechtan's well of lightning, whose waters rise up to

guard against the theft of the fiery source by his wife Bōand and drowns her in an estuary by the sea: "A deity hoards a fiery and effulgent power immersed in a body of water. His trust is challenged by one who is inherently unqualified to possess this treasure. . . . Three rounds of approach by the usurper result in three countermeasures, either retreats or attacks; in either instance, whether fleeing or pursuing, the advancing waters with their inherent fiery power create a watercourse or courses that after a worldwide circulation revert to their mythical source" (*Comparative Mythology*, 279).

7. Doniger, *The Rig Veda*, 108 and n. 109.

8. Puhvel, *Comparative Mythology*, 279.

9. Translated by Doniger in *Women, Androgynes, and Other Mythical Beasts*, 232.

10. *Padma Purāṇa* 5.18.159–98. Translated and discussed in Doniger, *Women, Androgynes, and Other Mythical Beasts*, 231.

11. From *Vāmana Purāṇa* 31.2–56, in *Classical Hindu Mythology: A Reader in the Sanskrit Purāṇas*, ed. and trans. Cornelia Dimmitt and J. A. B. van Buitenen (Philadelphia: Temple University Press, 1978), 185 ff.

12. For the interpretation of Śiva's role in the Amṛtamanthana myth as a shaman who sucks out evil substances from sick people, see David Smith, "The Dance of Siva," in *Perspectives on Indian Religion: Papers in Honour of Karel Werner*, ed. Peter Connolly, Bibliotheca Indo-Buddhica 30 (Delhi: Sri Satguru Publications, 1986), 93.

13. Jeanine Miller, "The Myth of the Churning of the Ocean of Milk," in *Perspectives on Indian Religion: Papers in Honour of Karel Werner*, ed. Peter Connolly, Bibliotheca Indo-Buddhica 30 (Delhi: Sri Satguru Publications, 1986), 70.

14. After its sequestering in Tibet at the Nghor monastery, the work became known to the West through the work of V. V. Ghokale and D. D. Kosmabi, and Daniel Ingalls's 1965 translation: *An Anthology of Sanskrit Court Poetry: Vidyākara's "Subhāṣitaratnakoṣa*," Harvard Oriental Series, vol. 44 (Cambridge: Harvard University Press, 1965).

15. How great shall we call Viṣṇu,
   in whose belly lies the universe;
   how great the serpent's hood where Viṣṇu lies?
   And yet the serpent lies but on a portion of the sea.
   (*Subhāṣitaratnakoṣa* 1209)

16. *Mahābārata*, Bk. 3, chaps. 99–103. See the edition of J. A. B. van Buitenen, trans. and ed. (Chicago: University of Chicago Press, 2004), 419–424.

17. Doniger, *Woman, Androgynes and Other Mythical Beasts*, 232: "The Indo-European mare was dangerous in her erotic powers precisely because

they were untamed; as raw forces of a Goddess, they were overpowering. For this reason, however, the mare was also able to bless and make fruitful, not merely because she was both mother and whore, both cow and mare, but because her powers flowed freely down to her mortal consort. The underwater mare, by contrast, is a symbol of angry, thwarted sexuality, of power blocked by authority."

18. Ibid., 232.

19. Ibid., 234–235.

20. Ibid., 235–236 *passim.*

21. Ibid., 237 (italics in original).

22. For example, Rajiv Malhotra, "The Wendy's Child Syndrome," Risa Lila 1, September 6, 2002; Internet-based critiques and responses published on the Sulekha website http://www.sulekha.com; "Does South Asian Studies Undermine India?" December 4, 2003, which "witnesses" recent alleged trends to "interpret Indian culture using Freudian theories to eroticize, denigrate and trivialize Indian spirituality."

23. I refer to the dissolution of all things into Viṣṇu as found in the *Viṣṇu Purāṇa* (6.3.14–41; 4.1–10), here cited from the versions in *Classical Hindu Mythology,* 42–43.

7. "Here End the Works of the Sea, the Works of Love"

1. *Mālatīmādhava* of Bhavabhūti 5.97.3.

2. The apocalyptic creature is divinely created, threatens the created order, can be contained only by the sea (no river is strong enough), and will be released at the end of time by divine power. "The underwater mare embodies the end, but under extreme control." Compare the parallel in "The Chapter of Jonah" in the eighth-century C.E. *Midrash Pirḳē de Rabbi Eliezer* (1916; repr., New York: Sepher-Hermon Press, 1981), in which the great fish that swallows Jonah brings him up alongside Leviathan and Jonah prophesies to the monster, "On thy account have I descended to see thy abode in the sea, for moreover, in the future will I descend and put a rope in thy tongue, and I will bring thee up and prepare thee for the feast of righteousness" (70). The Messianic feast of the righteous also figures in biblical and rabbinical texts (*b. Baba Bathra* 74a; *b. Ḥagigah* 14b, etc.) and the New Testament (Mt 26:29). Unlike the submarine mare, Leviathan is primordial, being one of the eight things that, according to Talmud, existed before the creation of the world.

3. Herman Melville, "The Pacific," in *Moby Dick,* 2nd ed., 150th anniversary ed., ed. Hershel Parker and Harrison Hayford (New York: Norton, 2002).

4. *Passio maximiani et Isaac donistratarum auctore macrobio,* Migne PL 8.772–773.

5. See the collection from the Celtic, Egyptian, Palestinian, and greater Levantine world in Helen Waddell, *Beasts and Saints,* ed. Esther de Waal (1934; repr., Grand Rapids, Mich.: Eerdmans, 1995), 16; and my discussion in "'He Who Sits in the Heavens Laughs': Recovering Animal Theology in the Abrahamic Traditions."

6. Maureen A. Tilley, "Martyrs, Monks, Insects, and Animals," in *An Ecology of the Spirit: Religious Reflection and Environment of Consciousness,* ed. Michael Barnes, *Annual Publication of the College Theology Society* 36 (1990): 98.

7. *Mādhyamakāvatāra* (Candakīrti's *The Entry into the Middle Way*), trans. C. W. Huntington Jr. with Geshé Namgyal Wangchen, from the Tibetan version by Pa tshab Nyi, in *The Emptiness of Emptiness: An Introduction to Early Indian Mādhyamika* (Honolulu: University of Hawaii Press, 1989), 151–152. Although as Huntington remarks, Candakīrti in his autocommentary does not remark upon this metaphor, there is a parallel in the Pali text *Cullavagga*: "And the Blessed One said to the Bhikkus: 'There are, O Bikkhus, in the great ocean, eight astonishing and curious qualities, by the constant perception of which the mighty creatures take delight in the great ocean. And what are the eight? O Bikkhus, the great ocean will not brook association with a dead corpse. Whatsoever dead corpse there be in the sea, that will it—and quickly—draw to the shore, and cast it out onto the dry ground. This is the third [of such qualities]'." See Huntington, *The Emptiness of Emptiness,* 222, n. 7, citing the translation of the *Cullavagga* by I. B. Horner (Oxford: Oxford University Press, 1952), 301.

8. See Adele Yarbro Collins, "The Combat Myth in the Book of Revelation," Harvard Dissertations in Religion, no. 9, ed. Caroline Bynum and George Rupp (Missoula, Montana: Scholars Press for *Harvard Theological Review,* 1976), 162.

9. Καὶ ἔδωκεν ἡ θάλασσα τοὺς νεκροὺς τοὺς ἐν αὐτῇ. Compare 1 Enoch 51:1: "In those days, Sheol will return all the deposits which she had received and hell will give back all which it owes. And he shall choose the righteous and the holy ones from among [the risen dead]"; and 1 Enoch 61: 5: "And these measurements shall reveal all the secrets of the depths of the earth, those who have been destroyed in the desert, those who have been eaten by the wild beasts, and those who have been eaten by the fish of the sea. So they all return and find hope in the day of the Elect One" See 1 *(Ethiopic Apocalypse of) Enoch,* trans. E. Isaac, in *Apocalyptic Literature and Testaments,* ed. James H. Charlesworth, The Old Testament Pseudepigrapha, vol. 1 (Garden City, N.Y.: Doubleday, 1983). Some have found the scenario of Revelations 20:13 to conflict with the events of 20:11: "Then I saw a great white throne and the one who sat on it; the earth and the heaven fled from his presence, and no place was found for them." Isn't the sea by this point

vanished, nonexistent? An early suggestion was that τὰ ταηεῖα, the "treasuries" in eschatological Jewish thought that stored the souls of the righteous, (thus natural analogues to the function of Death and Hades in the verse), were substituted for ἡ θάλασσα by the author of Revelations in order to introduce the doctrine of physical resurrection. See, for example, R. H. Charles, *A Critical and Exegetical Commentary on The Revelation of St. John*, vol. 2 (1920; repr., Edinburgh: T. & T. Clark, 1980), 194–197.

10. Wilfred J. Harrington, *Revelation*, Sacra Pagina Series, vol. 16, ed. Daniel J. Harrington (Collegeville, Minn.: Liturgical Press, 1993), 203.

11. Adele Yarbro Collins, *The Apocalypse, New Testament Message*, vol. 22, ed. Wilfred Harrington and Donald Senior (Wilmington, Del:. Michael Glazier, 1979), 144 (italics in original).

12. The ocean is often cast as supernatural player with agency at the end time, undoing and unleashing what had been put in place and contained to order the world. Of course, chaos is a characteristic of ocean itself. It is either slain and subdivided, as by Marduk in the *Enuma Elish*, "shut up" with brass bolts (as by God in the Psalms) or, most spectacularly, led on a hook in the Book of Job, where Leviathan and the sea seem interchangeable as world-threatening forces (see Jon D. Levenson, *Creation and the Persistence of Evil*). Consider Seneca's words: "There will come an age in the far-off years / When Ocean shall unloose the bonds of things, / When the whole broad earth shall be revealed."

13. *Gershon's Monster: A Story for the Jewish New Year*, retold by Eric A. Kimmel (New York, Scholastic Press, 2000).

14. Anne Wescott Dodd, *The Story of the Sea Glass* (Camden, Maine: Down East Books, 1999).

15. Richard Adams, *The Girl in a Swing* (New York: Knopf, 1980), 298.

16. "May you come to no sorrow through my death. I had no pity." The first sentence of the two Käthe utters is from "Der Zwerg" (The Dwarf), a poem by the nineteenth-century Viennese dramatist Matthäus von Collin.

17. On the contrast between the size of the sea and our lack of knowledge about it, see the 1998 *International Charter on Ocean Geography:* "Approaching the twenty-first century, humankind faces a paradox: the ocean covers more than two-thirds of the earth's surface and has a core role in providing living and non-living resources for the world's population, but it is much less understood than the terrestrial part of our planet." See Adalberto Vallega, *Sustainable Ocean Governance: A Geographical Perspective* (London: Routledge, 2001).

18. ClarkWolf, "Environmental Ethics and Marine Ecosystems, 24–27.

19. Shierry Weber Nicholsen, *The Love of Nature and the End of the World*, 153–154.

Bibliography

Adams, Richard. *The Girl in a Swing.* New York: Knopf, 1980.

Akima, Toshio. "The Songs of the Dead: Poetry, Drama, and Ancient Death Rituals of Japan," *Jounral of Asian Studies* 41 (May 1982): 485–509.

Allchin, Douglas. "The Poisoning of Minamata." SH*i*P*s* Teacher's Network, http://www1.umn.edu/ships/ethics/minamata.htm.

Allen, Douglas. *Myth and Religion in Mircea Eliade.* New York: Routledge, 2002.

An Anthology of Sanskrit Court Poetry: Vidyākara's "Subhāṣitaratnakoṣa," trans. David H. H. Ingalls. Harvard Oriental Series, vol. 44. Cambridge: Harvard University Press, 1965.

Baartmans, Frans. *Āpaḥ, the Sacred Waters: An Analysis of a Primordial Symbol in Hindu Myths.* Delhi: B.R. Publishing, 1990.

Bascom, Willard. *Waves and Beaches: The Dynamics of the Ocean Surface.* New York: Doubleday, 1964.

Boardman, John. *The Greeks Overseas: Their Early Colonies and Trade,* rev. ed. London: Thames & Hudson, 1999.

Boas, Franz. *The Central Eskimo.* 1888. Reprint, Lincoln: University of Nebraska Press, 1964.

———. "The Eskimo of Baffin Land and Hudson Bay." *Bulletin of the American Museum of Natural History* 15 (1901).

Bouthillier, Narelle. "She Down There: Sedna and Her Sisters of the Deep."
Unpublished paper, Harvard Divinity School, Cambridge, 2004.

Bremmer, Jan N. "Ino-Leucothea." In *The Oxford Classical Dictionary,* 3rd ed.,
ed. Simon Hornblower and Anthony Spawforth. Oxford: Oxford University
Press, 1996.

Brown, James, and Mahfuzuddin Ahmed. "Consumption and Trade of
Fish." Briefing paper no. 3, Conference of the Institute for European
Environmental Policy, "Sustainable EU Fisheries: Facing the Environmental
Challenges," Brussels, November 8–9, 2004.

Buchbaum, Robert. "Hiking to Georges Bank: What If You Could Walk Under
the Gulf of Maine?" *Sanctuary* 41, no. 6 (2002): 3–9.

Burkert, Walter. *Greek Religion.* Cambridge: Harvard University Press, 1987.

———. *Homo Necans: The Anthropology of Ancient Greek Sacrificial Ritual and
Myth,* trans. Peter Bing. Berkeley: University of California Press, 1983.

Cains, John. "Waterway Recovery." *Water Spectrum* (Fall 1978).

Carson, Rachel. *The Sea Around Us.* 1950. Reprint, New York: Oxford University
Press, 1989.

———. *Under the Sea Wind.* 1941. Reprint, New York: Penguin Books, 1996.

Classical Hindu Mythology: A Reader in the Sanskrit Purāṇas, ed. Cornelia Dimmitt and J. A. B. van Buitenen. Philadelphia: Temple University Press, 1978.

Coleman, Loren, and Patrick Hughe. *The Field Guide to Lake Monsters, Sea
Serpents, and Other Mystery Denizens of the Deep.* New York: Penguin, 2003.

Collins, Adele Yarbro. *The Apocalypse. New Testament Message,* vol. 22, ed. Wilfrid Harrington and Donald Senior. Wilmington, Delaware: Michael Glazier,
1979.

———. "The Combat Myth in the Book of Revelation." Harvard Dissertations
in Religion, no. 9, ed. Caroline Bynum and George Rupp. Missoula, Montana: Scholars Press for *Harvard Theological Review,* 1976.

Commission on Physical Sciences, Mathematics, and Applications (CPSMA).
Disposal of Industrial and Domestic Wastes: Land and Sea Alternatives.
Washington, D.C.: National Academy Press, 1984.

Cramer, Deborah. *Great Waters: An Atlantic Passage.* New York: Norton, 2001.

Day, John. *God's Conflict with the Dragon and the Sea.* Cambridge: Cambridge
University Press, 1985.

Delivorrias, Angelos, ed. *Greece and the Sea.* Exhibition catalog, Amsterdam,
De Nieuwe Kerk, October 29–December 10, 1987. Athens: Ministry of Culture and Benaki Museum, 1987.

Diogenes, Laertius, *Pathagoras,* trans. R. D. Hicks. Cambridge: Harvard University Press, 1958.

Dodd, Anne Wescott. *The Story of the Sea Glass*. Camden, Maine: Down East Books, 1999.

Douglas, Mary. *Purity and Danger: An Analysis of the Concept of Pollution and Taboo*. 1964. Reprint, London: Routledge, 2004.

———. *Natural Symbols: Explorations in Cosmology*. 1970. Revised edition, New York: Routledge, 1996.

———. *"Purity and Danger* Revisited." Lecture, Institution of Education, The University of London, May 12, 1980. Published in *London Times Literary Supplement*, September 19, 1980, 1045–1046.

———, and Aaron Wildavsky. *Risk and Culture: An Essay on the Selection of Technical and Environmental Dangers*. Berkeley: University of California Press, 1983.

———. "Sacred Contagion." In *Reading Leviticus: A Conversation with Mary Douglas*, ed. John F. A. Sawyer. Journal for the Study of the Old Testament Supplement Series 227, 86–106. Sheffield: Sheffield Academic Press, 1996.

Dronke, Peter, and Ursula Dronke. "Growth of Literature: The Sea and the God of the Sea." H. M. Chadwick Memorial Lecture. Cambridge: Department of Anglo-Saxon, Norse and Celtic, Cambridge University, 1998.

Eck, Diana L. *Banaras, City of Light*. 1982. Reprint, New York: Columbia University Press, 1999.

Edsman, Carl-Martin. "Boats," trans. Kjersti Board. In *The Encyclopedia of Religion*, ed. Mircea Eliade. New York: Macmillan, 1987.

Eliade, Mircea. *The Sacred and the Profane: The Nature of Religion*, trans. Willard R. Trask. New York: Harcourt, Brace and World, 1959.

———. *Patterns in Comparative Religion*, trans. Rosemary Sheed. 1958. Reprint, Lincoln: University of Nebraska Press, 1966.

———. *Cosmos and History: The Myth of the Eternal Return*. 1959. Reprint, New York: Garland, 1985.

Ellis, Richard. *The Empty Ocean: Plundering the World's Marine Life*. Washington, D.C.: Island Press/Shearwater, 2003.

1 (*Ethiopic Apocalypse of*) *Enoch*, trans. E. Isaac. In *The Old Testament Pseudepigrapha*, vol. 1, *Apocalyptic Literature and Testaments*, ed. James H. Charlesworth. Garden City, N.Y.: Doubleday, 1983.

Euripides. *Iphigenia in Tauris,* ed. Maurice Platnauer. 1938. Reprint, Bristol: Bristol Classical Press, 1984.

———. *Iphigenia in Tauris*, trans. Witter Bynner. In The Complete Greek Tragedies, *Euripides II*, ed. David Grene and Richard Lattimore. Chicago: University of Chicago Press, 1956.

Fardon, Richard. *Mary Douglas: An Intellectual Biography.* London: Routledge, 1999.

Faulkner, William. *Thinking of Home: William Faulkner's Letters to His Mother and Father,* 1918–1925. New York: Norton, 1991.

Feely, Richard A., et al. "The Impact of Anthropogenic CO_2 on the $CaCO_3$ System in the Oceans." *Science* 16 (July 2004): 362–366.

Fienup-Rirordan, Anne, ed., and Marie Meade, trans. *Agayuliyararput: Kegginaqutm Kangiit-llu* [*Our Way of Making Prayer: Yup'ik Masks and the Stories They Tell*]. Seattle: Anchorage Museum of History and Art in association with University of Washington Press, 1996.

Gershon's Monster: A Story for the Jewish New Year. Retold by Eric A. Kimmel. New York: Scholastic Press, 2000.

Goldberg, Edward D. "The Oceans as Waste Space." In *Impact of Marine Pollution on Society,* ed. Virginia K. Tippe and Dana R. Kester. Center for Ocean Management Studies, The University of Rhode Island. South Hadley, Mass.: J. F. Bergin, 1982.

Gwynn, Richard. *Way of the Sea: The Use and Abuse of the Oceans.* Bideford Devon: Green Books, 1987.

Hamilton-Paterson, James. *Seven-Tenths: The Sea and Its Thresholds.* London: Hutchinson, 1992.

Harrington, Wilfred J. *Revelation.* Sacra Pagina Series, vol. 16, ed. Daniel J. Harrington. Collegeville, Minn.: Liturgical Press, 1993.

Hazra, Rajendra Chandra. "Problems Relating to Śivapurāṇa." In *Our Heritage, Bulletin of the Department of Postgraduate Training and Research* (Sanskrit College, Calcutta) 1, no. 1 (1953): 65.

Heidel, Alexander. *The Babylonian Genesis: The Story of Creation.* 1942. Reprint, Chicago: The University of Chicago Press, 1951.

Helvarg, David. *Blue Frontier: Saving America's Living Seas.* New York: Freeman, 2001.

Henrichs, Albert. "'Thou Shalt Not Kill a Tree': Greek, Manichaean and Indian Tales," *Bulletin of the American Society of Papyrologists* 16 (1979): 85–108.

Hetherington, Kevin. "Secondhandness: Consumption, Disposal, and Absent Presence." *Environment and Planning D: Society and Space* 22 (2004): 157–173.

Heyerdahl, Thor. *The Ra Expeditions.* Garden City, N.Y.: Doubleday, 1971.

Holtved, Erik. "The Eskimo Myth About the Sea-Woman: A Folkloristic Sketch." *Folk* 8/9 (1966–1967): 145–153.

———. *The Polar Eskimos: Language and Folklore.* Meddelelser om Grønland 152, nos. 1 and 2. Copenhagen: C. A. Reitzel, 1951.

Homer. *The Odyssey of Homer*, trans. Richmond Lattimore. New York: Harper-Perennial, 1991.

Hultkrantz, Åke. "The Religion of the Goddess in North America." In *The Book of the Goddess, Past and Present: An Introduction to Her Religion*, ed. Carl Olson. New York: Crossroad, 1983.

———. *The Religions of the American Indians*, trans. Monica Setterwell. Berkeley: University of California Press, 1979.

Jātaka Tales, trans. "various hands," vol. 3, no. 316, ed. E. B. Cowell. Cambridge: Cambridge University Press, 1895–1907.

Jenness, Diamond. *The Life of the Copper Eskimos. Report of the Canadian Arctic Expedition 1913–18*, vol. 12. Ottawa: F. C. Acland, 1922.

Keller, Catherine. *Face of the Deep: A Theology of Becoming*. London: Routledge, 2003.

Kleiven, Inge. *Mitârtut: Vestiges of the Eskimo Sea-Woman Cult in West Greenland.* Meddelelser om Grønland 161, no. 5. Copenhagen: C. A. Rietzel, 1960.

———. "Sedna." In *The Encyclopedia of Religion*, ed. Mircea Eliade. New York: Macmillan, 1987.

Kowalzig, Barbara. "From Drama to Ritual: Performances of, by, and for Dionysos." In *From Ritual to Theater: The Prehistory of Theatre in Ancient Greece and Elsewhere*, ed. Margaret Miller and Eric Csapo. Forthcoming. Cambridge: Cambridge University Press, 2007.

Kroeber, A. L. "Tales of the Smith Sound Eskimo." *Journal of American Folk-lore* 12 (1899).

Lauterbuch, Jacob Z. "Tashlik: A Study in Jewish Ceremonies." *Hebrew Union College Annual* 11 (1936): 207–340.

Lenssen, Nicholas. "The Ocean Blues." In *The World Watch Reader on Global Environmental Issues*, ed. Lester R. Brown. New York: Norton, 1991.

Leopold, Aldo. *A Sand County Almanac*. New York: Oxford University Press, 1966.

Levenson, Jon D. *Creation and the Persistence of Evil: The Jewish Drama of Divine Omnipotence*. New York: Harper and Row, 1985.

———. "Exodus and Liberation." In Jon Levenson, *The Old Testament, the Hebrew Bible, and Historical Criticism: Jews and Christians in Biblical Studies*, 127–159. Louisville: Westminster/John Knox, 1993.

Lloyd-Jones, Hugh. "Ritual and Tragedy." In *Ansichten griechischer Rituale. Geburtstags-Symposium für Walter Burkert*, ed. Fritz Graf. Stuttgart: Teubner, 1998.

Malamat, Abraham. "The Sacred Sea." In *Sacred Space: Shrine, City, Land*, ed. Benjamin Kedar and R. J. Zwi Werblowsky. New York: New York University Press, 1998.

Martel, Yann. *Life of Pi*. New York: Harcourt, 2001.

Melville, Herman. *Moby Dick*. 2nd ed., 150th anniversary ed., ed. Hershel Parker and Harrison Hayford. New York: Norton, 2002.

Merkur, Daniel. *Becoming Half-Hidden: Shamanism and Initiation Among the Inuit*. Stockholm: Amqvist and Wiksell, 1985.

———. *Powers Which We Do Not Know: The Gods and Spirits of the Inuit*. Moscow: The University of Idaho Press, 1991.

Miller, Jeanine. "The Myth of the Churning of the Ocean of Milk." In *Perspectives on Indian Religion: Papers in Honour of Karel Werner*, ed. Peter Connolly. Bibliotheca Indo-Buddhica 30. Delhi: Sri Satguru Publications, 1986.

Mitchell, John H. "The Go-Between." *Sanctuary* 41, no. 6 (2002): 2.

Myths from Mesopotamia, trans. and ed. Stephanie Dalley, rev. ed. Oxford: Oxford University Press, 2000.

Nagy, Gregory. *The Best of the Achaeans: Concepts of the Hero in Archaic Greek Poetry*. Baltimore: The Johns Hopkins University Press, 1979.

Neusner, Jacob. *Tractate Miqvaot*. In Jacob Neusner, *Halakhah: Encyclopedia of the Law of Judaism*. Leiden: Brill, 1999.

The New Oxford Annotated Bible with the Apocrypha, ed. Bruce Metzger and Roland Murphy. New York: Oxford University Press, 1994.

Nicholsen, Shierry Weber. *The Love of Nature and the End of the World: The Unspoken Dimensions of Environmental Concern*. Cambridge: MIT Press, 2002.

Norman, Howard, ed. *Traditional Stories of Eskimo and Indian People*. New York: Pantheon Books, 1990.

O'Flaherty, Wendy Doniger. "The Submarine Mare in the Mythology of Śiva." *Journal of the Royal Asiatic Society of Great Britain and Ireland* 1 (1971): 9–27.

———, trans. *Hindu Myths: A Sourcebook Translated from the Sanskrit*. Harmondsworth: Penguin, 1975.

———. *Women, Androgynes, and Other Beasts*. Chicago: University of Chicago Press, 1980.

———, trans. *The Rig Veda: An Anthology*. Harmondsworth: Penguin, 1981.

Oliver, Mary. "The Sea." In *American Primitive: Poems*, 69-70. Boston: Little, Brown, 1983.

O'Meara, John, trans. *The Voyage of Saint Brendan: Journey to the Promised Land* (*Navigatio Sancti Brendani Abbatis*). Gerrards Cross: Colin Smythe, 1991.

Oosten, Jarich G. "The Symbolism of the Body in Inuit Culture." In *Visible Religion I: Commemorative Figures*, ed. H. G. Kippenberg. Leiden: Brill, 1982.

Pa tshb Nyi. *Mādhyamakāvatāra* (Candakīrti's *The Entry into the Middle Way*), trans. C. W. Huntington Jr., with Geshé Namgyal Wangchen. In *The Emptiness of Emptiness: An Introduction to Early Indian Mādhyamika*, 149-196. Honolulu: University of Hawaii Press, 1989.

Parker, Robert. *Miasma: Pollution and Purification in Early Greek Religion.* New York: Oxford University Press, 1990.

Patton, Kimberley C. "From Ritual to Drama: A Comparative Analysis." In *From Ritual to Theater: The Prehistory of Theatre in Ancient Greece and Elsewhere*, ed. Margaret Miller and Eric Csapo. Forthcoming. Cambridge: Cambridge University Press, 2007.

———. "'He Who Sits in the Heavens Laughs': Recovering Animal Theology in the Abrahamic Traditions." *Harvard Theological Review* 93, no. 4 (2000): 401–434.

———. "'The Sea Can Wash Away All Evils': Modern Marine Pollution and the Ancient Cathartic Ocean." Paper delivered at the Annual Meeting of the American Academy of Religion, San Francisco, November 23, 1992.

Pausanias. *Description of Greece*, trans. W. H. S. Jones. Cambridge: Harvard University Press, 1918–1935.

———. *Description of Greece*, trans. Peter Levi. Harmondsworth: Penguin, 1971.

Pelly, David. *Sacred Hunt: A Portrait of the Relationship Between Seals and Inuit.* Vancouver: Greystone Books, and Seattle: University of Washington Press, 2001.

Pew Oceans Commission. *America's Living Oceans: Charting a Course for Sea Change.* Arlington, Va.: May 2003.

Pickstock, Catherine. *After Writing: On the Liturgical Consummation of Philosophy.* Oxford: Blackwell, 1998.

Potter, J. R. and M. A. Chitre. "Ambient Noise: Imaging in Warm Shallow Seas; Second-order Moment and Model-based Imaging Algorithms." *Journal of the Acoustical Society of America* 106 (1999): 3201–3210.

Puhvel, Jaan. *Comparative Mythology.* Baltimore: Johns Hopkins University Press, 1987.

Qitsualik, Rachel Attituq. "The Problem with Sedna." Six-part series. *Nunatsiaq News*, March/April 1999.

Raban, Jonathan, ed. *The Oxford Book of the Sea.* New York: Oxford University Press, 1992.

Rao, Velcheru Narayana. "Purāṇa as Brahmanic Ideology." In *Purāṇa Perennis: Reciprocity and Transformation in Hindu and Jaina Texts*, ed. Wendy Doniger. Albany: State University of New York Press, 1993.

Rasmussen, Knud. *Across Arctic America: Narrative of the Fifth Thule Expedition.* 1927. Reprint: Fairbanks: The University of Alaska Press, 1999.

————. *Intellectual Culture of the Igulik Eskimos. Report of the Fifth Thule Expedition 1921–1924,* vol. 7, no. 1. 1929. Reprint, New York: AMS Press, 1976.

————. *Intellectual Culture of the Igulik Eskimos. Report of the Fifth Thule Expedition 1921–1924,* vol. 7, no. 2. 1930. Reprint, New York, AMS Press, 1976.

————. *The Netsilik Eskimos: Social Life and Spiritual Culture. Report of the Fifth Thule Expedition 1921–1924,* vol. 8, nos. 1 and 2. Copenhagen: Gyldendal, 1931.

————. *Intellectual Culture of the Copper Eskimos. Report of the Fifth Thule Expedition 1921–1924,* vol. 9. Copenhagen: Gyldendal, 1932.

Rennie, Bryan, ed. *Changing Worlds: The Meaning and End of Mircea Eliade.* Albany: State University of New York Press, 2001.

Rink, Henry. *Tales and Traditions of the Eskimo.* Edinburgh: William Blackwood and Sons, 1875.

Rocher, Ludo. *The Purāṇas. A History of Indian Literature,* vol. 2, part 3, ed. Jan Gonda. Wiesbaden: Harrassowitz, 1986.

Rudhart, Jean. "Deities of Water in Greek Mythology." In *Mythologies,* ed. Yves Bonnefoy, 379-384. Chicago: The University of Chicago Press, 1991.

Ryan, William, and Walter Pitman. *Noah's Flood: The New Scientific Discoveries About the Event That Changed History.* New York: Simon & Schuster, 1998.

Sabine, Christopher L., et al. "The Oceanic Sink for Anthropogenic CO_2." *Science* 16 (July 2004): 367–371.

Schimmel, Annemarie. *And Muhammad Is His Messenger: The Veneration of the Prophet in Islamic Piety.* Chapel Hill: University of North Carolina Press, 1985.

Schulz, Raimund. *Die Antike und das Meer.* Darmstadt: Primus Verlag, 2005.

Schwenk, Theodor. *Water: The Element of Life,* trans. Marjorie Spock. New York: Anthroposophic Press, 1989.

Seidelman, Harold, and James Turner. *The Inuit Imagination: Arctic Myth and Sculpture.* New York: Thames & Hudson, 1994.

Shields, Rob. "Spatial Stress and Resistance: Social Meanings of Spatialization." In *Space and Social Theory: Interpreting Modernity and Postmodernity,* ed. Georges Benko and Ulf Strohmayer. Oxford: Blackwell, 1997.

Śiva Purāṇa, trans. a board of scholars. Delhi: Motilal Banarsidass, 1973/1974.

Smith, David. "The Dance of Śiva." In *Perspectives on Indian Religion: Papers in Honour of Karel Werner,* ed. Peter Connolly. Bibliotheca Indo-Buddhica 30. Delhi: Sri Satguru Publications, 1986.

Smith, Jonathan Z. *Map Is Not Territory: Studies in the History of Religions.* Chicago: The University of Chicago Press, 1993.

Søby, Regitze Margrethe. "The Eskimo Animal Cult." *Folk (Dansk Etnografisk Tidsskrift)*: Essays Presented to Erik Holtved on His Seventieth Birthday 11/12 (1969/70): 43–78.

Sokolowski, Franciszek. *Lois sacrées des cités grecques (LSCG).* Paris: Éditions de Boccard, 1969.

Sonne, Birgitte. "The Acculturative Role of Sea Woman." *Meddeleser om Grønland, Man and Society* 13 (1990): 3–34.

Sophocles. *Ajax*, trans. John Moore. In The Complete Greek Tragedies, *Sophocles II*, ed. David Grene and Richmond Lattimore. Chicago: University of Chicago Press, 1957.

———. *Ajax*, ed. W. B. Stanford. 1963. Repr., Bristol: Bristol Classical Press, 1981.

———. *Oedipus the King*, trans. David Grene. In The Complete Greek Tragedies, *Sophocles I*, ed. David Grene and Richmond Lattimore. Chicago: University of Chicago Press, 1954.

———. *Oedipus Rex.* Ed. R. D. Dawe. Cambridge Greek and Latin Classics. Cambridge: Cambridge University Press, 1982.

Souvorov, Aleksandr V. *Marine Ecologonomics: The Ecology and Economics of Marine Natural Resources Management.* Developments in Environmental Economics, vol. 6. Amsterdam: Elsevier, 1999.

Specter, Michael. "The World's Oceans Are Sending an S.O.S." *New York Times,* May 3, 1992.

Staal, Frits. *Rules Without Meaning: Ritual, Mantras, and the Human Sciences.* New York: Peter Lang, 1989.

Stafford, D. M. *Tangata Whenna: The World of the Maori.* Auckland: Reed Books, 1996.

Staudigel, Hubert, et al., "Vailulu'u Seamount, Samoa: Life and death on an active submarine volcano," *Proceedings of the National Academy of Sciences* 103: 17 (April 25, 2006); 6448–6453.

Stocker, Michael. "Fish, Mollusks, and Other Sea Animals' Use of Sound, and the Impact of Anthropogenic Noise in the Marine Acoustic Environment." Unpublished paper, Earth Island Institute, San Francisco, 2002.

Stow, Dorrick. *Oceans: An Illustrated Reference.* Chicago: University of Chicago Press, 2005.

Strathern, Marilyn. "The Aesthetics of Substance." In Marilyn Strathern, *Property, Substance and Effect : Anthropological Essays on Persons and Things,* 45–64. London and New Brunswick, NJ: Athlone Press, 1999.

Swinton, Nelda. *La Déese inuite de la mer/The Inuit Sea Goddess.* Montreal: Musée des beaux-arts de Montréal, 1980.

Suzuki, David, with Amanda McConnell. *The Sacred Balance: Rediscovering Our Place in Nature.* Amherst, N.Y.: Prometheus Books, 1998.

Taube, Karl. *Aztec and Maya Myths,* 2nd. ed. Austin: University of Texas Press, 1995.

Tilley, Maureen A. "Martyrs, Monks, Insects, and Animals." In *An Ecology of the Spirit: Religious Reflection and Environment of Consciousness,* ed. Michael Barnes. *Annual Publication of the College Theology Society* 36 (1990): 97–109.

Tolkien, J. R. R. *The Lord of the Rings.* 1955. Reprint, Boston: Houghton Mifflin, 1994.

Trainer, Kevin. *Relics, Ritual, and Representation in Buddhism: Rematerializing the Sri Lanka Peravada Tradition.* Cambridge: Cambridge University Press, 1997.

U.S. Commission on Ocean Policy. "An Ocean Blueprint for the 21st Century." Report delivered to the President and Congress on September 20, 2004. *http://www.oceancommission.gov/.*

Vallega, Adalberto. *Sustainable Ocean Governance: A Geographical Perspective.* London: Routledge, 2001.

van Buitenen, J. A. B., trans. and ed. *Mahābārata.* 1973. Reprint, Chicago: University of Chicago Press, 2004.

Vernant, Jean-Pierre. "Greek Cosmogonic Myths." In *Mythologies,* ed. Yves Bonnefoy, 366-375. Chicago: University of Chicago Press, 1991.

Wachsmuth, Dietrich. *ΠΟΜΠΙΜΟΣ Ο ΔΑΙΜΟΝ: Untersuchung zu den antiken Sakralhandlungen bei Seereisen.* Berlin: Ernst-Reuter-Gesellschaft, 1967.

Waldeman, Amy, and David Rohde. "Fearing a Sea That Once Sustained, Then Killed." *New York Times,* January 5, 2005.

Wilson, T. Roger, and J. Dennis Burton. "Why Is the Sea Salty? What Controls the Composition of Ocean Water?" In *Understanding the Oceans: A Century of Ocean Exploration,* ed. Margaret Deacon, Tony Rice, and Colin Summerhayes. London: UCL Press, 2001.

Wolf, Clark. "Environmental Ethics and Marine Ecosystems: From a 'Land Ethic' to a 'Sea Ethic.'" In *Values at Sea: Ethics for the Marine Environment,* ed. Dorinda Dallmeyer. Athens: University of Georgia Press, 2003.

Woodward, Colin. *Ocean's End: Travels Through Endangered Seas.* New York: Basic Books, 2000.

Yiannis, Gabriel. "Organization Nostalgia: Reflections on the Golden Age." In *Emotion in Organizations,* ed. Stephen Fineman. London: Sage, 1993.

Zuesse, Evan. "Ritual." In *The Encyclopedia of Religion,* ed. Mircea Eliade. New York: Macmillan, 1987.

———. *Ritual Cosmos: The Sanctification of Life in African Religions.* Athens: Ohio University Press, 1979.